78th Annual International Association of Railway Operating Officers Meeting 2014

AA005344

Montreal, Canada
21-23 September 2014

ISBN: 978-1-5108-2225-2

Printed from e-media with permission by:

Curran Associates, Inc.
57 Morehouse Lane
Red Hook, NY 12571

Some format issues inherent in the e-media version may also appear in this print version.

Printed by Curran Associates, Inc. (2016)

For permission requests, please contact International Association of Railway Operating Officers Inc. (IAROO) at the address below.

International Association of Railway Operating Officers Inc. (IAROO)
22174 N. Reinbold Dr.
Maricopa, AZ 85138
USA

Phone: (406) 370-6977

www.iaroo.org

Additional copies of this publication are available from:

Curran Associates, Inc.
57 Morehouse Lane
Red Hook, NY 12571 USA
Phone: 845-758-0400
Fax: 845-758-2633
Email: curran@proceedings.com
Web: www.proceedings.com

Seventy Eighth Annual

Proceedings
of
The International
Association of Railway
Operating Officers
2014

Purpose of the Association

The object of this Association shall be to provide an organization for the improvement of its members in an educational way and advancement of the interest of the railroads through the adoption of the best methods of improving Locomotive Operation and Train Handling.

Our meetings being open to impartial discussions of subjects, the remarks contained herein are therefore not necessarily the opinions of this association, but are the free and independent expressions of the individual.

The International Association of Railway Operating Officers
22174 N Reinbold Dr.
Maricopa, AZ 85138

2014 Officers

President:
Randy Eardensohn

MEET THE PRESIDENT

IAROO'S 2013-2014 President is Randy Eardensohn. Randy began his career with the Union Pacific Railroad in 1979 as a Switchman/Brakeman in Las Vegas, Nevada. He was promoted to Locomotive Engineer in 1981 and worked in California, Nevada, Utah, Oregon and Idaho in that capacity.

In 1992 he was promoted to Manager of Operating Practices and worked as MOP in Arkansas, Oklahoma and Kansas until 2006, when he joined the system Operating Practices team in Omaha.

In 2012 he was promoted to Director of Regional Operations for the Northern Region of the Union Pacific. In his current position he is responsible for Safety, Train Handling, Operating Rules and Regulatory compliance in Wyoming, Colorado, Kansas, Nebraska, Minnesota, Iowa, Illinois and Missouri.

Randy and his wife of 28 years, Renee, have four children; Emily (23), Nick (22) LeeAnne (20) and Matt (17). When his many career responsibilities allow he enjoys golf, fishing, and watching baseball.

Vice Presidents

1st Vice President
Warren Egan,
TUV-RSI

2nd Vice President
Aaron Ratledge,
BNSF

3rd Vice President
Foster Peterson,
Full Service Rail

Photo not available

3rd Vice President
Mike Mattson,
Montana Rail Link

Executive Committee:

Darrel Papproth, Paducah & Louisville RR
Foster Peterson, Full Service Rail
John Punwani, FRA
Mike Mattson, Montana Rail Link
Robert Repola, BNSF
John Heater, AKRR

Advisory Committee:

Dwight Anderson, UPRR (retired)
Tony Crabb, RS&TR
W.J. Egan, CSX (retired)
W.C. Faulhaber, NS
M.L. Volkmar, UP (retired)
J.C. Kissinger, CSX (retired)
J.J. Nadzam Conrail (retired)
T.S. Pape, Amtrak
J.M. Quilty, BNSF (retired)
M.F. Schmitt,
Gary Wolf, Wolf Railway Consulting
Bill Barrington, BNSF
Frank Hernandez, BNSF
Larry Breeden, TCS

Regional Consultants:

Keith Jensen, UPRR
Art Regenold, NS
Casey Calkins, MRL
Dennis Merrell, CSX
Don Marson, Pine Tree Consulting
Gregg Cooney, MRL
Jeff Gird, UPRR
Nick Mizelle, BNSF
Steve Salmons, CSX
Darrell Kreuger, BNSF
Taylor James, CSX
Andy John, CSX

PAST PRESIDENTS

Anderson, D. D.	Union Pacific
Arakalien, M. X.	Chicago and Northwestern
Baker, M. M.	Union Pacific
Barrington, B.	BNSF
Breeden, L.	Union Pacific
Broom, T. R.	Florida East Coast
Buffalow, R. G.	Denver & Rio Grande Western
Crabb, T.	RS&TR
Eck, H. C.	Chessie
Egan, W. J.	CSXT
Faulhaber, W. C.	NS
Finnerty, L. G.	CN
Isham, R.	Union Pacific
Hagemo, T.	MRL
Hernandez, F.	BNSF
Kautzman, K.A.	MRL
Kissinger, J. C.	CSX Trans.
Lawrence, L. S.	ATSF
Mauney, I. Ray	Southern
Mecaskey, J. R.	ATSF
Menefee, D.	CSX
Nadzam, J. J.	Conrail
Pape, T. S.	AMTRAK
Pryke, B.	AKRR
Quilty, J. M.	BNSF
Roberts, Jack	Chessie
Sanders, B. D.	Kansas City Southern
Schmitt, M. F.	WC
Sparks, F. L.	ATSF
Tindol, C. J.	Grand Trunk Western
Volkmar, M. L.	Union Pacific
Wolf, G.	Rail Sciences, Inc.
Yerkes, D. R.	Conrail

Index

BNSF 90 CFM Pilot Test Waiver........................ 9

LOCOTROL DP over Cellular Communications...... 25

Prevention of Yard, Terminal & Industry Derailments 39

Investigating for Organizational and Management Factors 111

Natural Gas Fuel Tender Performance Standard 127

PST Rail Operations Simulator 145

Investigating Distributed Power Failures 155

Implementing Top of Rail Mobile Technology....... 167

Trip Optimizer, Energy Management and Consist Control 185

Coupler Force Environment and Coupler Durability 207

Longitudinal vs. Lateral Train Force and Managing Wheel/Rail Interaction Through Train Operations.. 235

Steve Zuiderveen, FRA

Steve Zuiderveen comes to us from the Motive Power and Equipment Division of the Federal Railroad Administration Headquarters in Washington, DC. As a Railroad Safety Specialist his subject matter portfolio includes rear-end maker lights, blue signal, power brakes and securement. He also assists in the areas of tourists, historic railroads, and steam locomotives.

Steve began his railroad career as a steam locomotive fireman for Genesee County in Flint, Michigan, and worked for a passenger equipment operator in Wisconsin before moving to Scranton, Pennsylvania, where he was the passenger car foreman for the Steamtown Foundation. At that time he also formed the DL&W Car Company, which until his present engagement with the FRA provided contract railroad passenger car services to its clients. Since Steamtown, Steve has been a Project Manager for the Maryland MARC Train Service for equipment overhauls and a Staff Engineer for ENSCO, Inc., where he managed the construction and maintenance of track geometry cars for the FRA. Steve has been with the FRA for the past six years and lives with his wife Laura in Sykesville, Maryland.

Jeff Garrels, BNSF

Jeff Garrels comes to us from the Operating Practices team within the Safety and Technical Training Department at BNSF Railway headquartered in Fort Worth, Texas where his current position is the Director of Train Handling. In his current role, he provides guidance on train handling and makeup rules, locomotive event recorder analysis, and integrated locomotive cab technology. Currently Jeff is the acting chairman for BNSF's system locomotive cab committee and the AAR Cab Standardization Executive Committee.

Jeff began his railroad career with the Santa Fe Railroad in 1995, working in the craft as a conductor/brakeman/switchman in Barstow, California. Shortly thereafter, he promoted to a locomotive engineer, working various routes on BNSF's southern trans-con. Jeff continued to expand upon his railroad background by serving as a company officer in various capacities including Trainmaster, Road Foreman of Engines, Manager Corridor Operations, Locomotive Engineer Instructor, and Senior Manager Locomotive Engineer Training at BNSF's Technical Training Center. Prior to his career in the railroad, her served in the U.S. Navy within the submarine fleet as an electrical nuclear power operator. Jeff currently lives with his wife in Fort Worth, Texas.

90 CFM - DP Operations

Present-day FRA and previous Transport Canada Air Brake Rules

- Limit the air flow not to exceed 60 CFM for an entire train
- No more than 15 psi brake pipe pressure gradient between head-end and rear-end.

These requirements have been in place since 1984 in Canada and 1989 in the U.S.

90 CFM - DP Operations

Why the need?

- Specifically for cold weather operations
- Northern Hi-Line – Minneapolis to Whitefish, MT
- BNSF's Powder River Division

Should DP operations continue to be restricted?

- One advantage of Distributed Power (DP) is the ability to charge and maintain brake pipe by multiple air sources, distributed throughout the train.

BNSF RAILWAY

6/5/2015 3

90 CFM - DP Operations

Canadian Pacific Efforts

- "The extensive series of tests, that began in 2009, validated our theory that air flow rates can be increased from today's maximum of 60 CFM to reach as high as 90 CFM on Distributed Power trains.

All of our testing proved that this increase in train leakage:

- Did not adversely affect LOCOTROL® Communication Interruption recovery protocols.
- Did not have any negative impacts on overall air brake functionality of train brakes under certain conditions of Distributed Power and train segment lengths."

BNSF RAILWAY

6/5/2015 4

90 CFM - DP Operations

Canadian Pacific Efforts

- "On August 9th, 2011, the Railway Association of Canada (RAC) filed with the Minister Of Transport, a set of proposed revisions to the Railway Freight & Passenger Train Brake Inspection and Safety Rules.
- This filing was to amend the rules to allow for the operation of freight trains with multiple sources of air, to operate with combined air flows not to exceed 90 CFM in total with no individual air source having a flow greater than 60 CFM.
- On November 21st, 2011, the Minister approved the revisions to the Rule"

BNSF RAILWAY

6/5/2015 5

BNSF Testing – Great Falls, MT

August 29, 2012

- Participants
  - FRA, BNSF, NS, CPR
  - Train Make-up - 101 grain hoppers – 2x2 DP configuration

BNSF RAILWAY

6/5/2015 6

BNSF Testing – Great Falls, MT

Conclusions

- Brake applications and releases ranging from a minimum reduction to a full service application were made without any problems at brake pipe leakage rates of 60 CFM and 90 CFM.
- No significant differences were noted between the brake applications made at 90 CFM versus the brake applications made at 60 CFM.
- These results support the work previously performed by the Canadian Pacific.
- November 15,2012 - BNSF petitions for waiver of 60 CFM limit for DP trains.
- On March 26, 2013, the Board granted BNSF a Pilot Test Waiver for the use of 90 CFM AFM on DP trains.

BNSF RAILWAY

6/5/2015 7

FRA-2012-0091

- On March 26, 2013, the Board granted BNSF a Pilot Test Waiver for the use of 90 cfm AFM on DP trains
- Pilot waivers are limited to one railroad until sufficient supporting data is generated
- The Board preferred the use of a Test Waiver to determine if Canadian practice was appropriate for U.S. regulatory and operating environment

Federal Railroad Administration

8

Test Waiver Summary

1. This pilot test waiver only applies to DP trains operated by BNSF, with AFM readings between 60 and 90 cfm
2. A test committee oversees this pilot test waiver. Members include:
   - Brake and/or DP manufacturers, Rail Labor, an FRA test monitor, Railroad Representatives, AAR brake committee members, and others involved with the operation of DP trains at higher CFM airflows.
3. The test committee is responsible for the test plan, including testing metrics and parameters.
4. All testing procedures and parameters are subject to the approval of the FRA test monitor.

Federal Railroad Administration

9

Test Waiver Summary

5. FRA retains the right to modify or suspend any test planned under the test plan.
6. BNSF provides a monthly report of all trains.
7. The test committee shall meet every 6 months:
   - Reviews the test data
   - Votes to continue the program for an additional 6 months.
8. BNSF ensures that the AFM indicator on test train locomotives are in proper calibration

Federal Railroad Administration

10

Test Waiver Summary

9. BNSF reports immediately to FRA any accident/incident/ injury to any railroad employee that involves test trains
10. Trains operated under this pilot test waiver operate under the operating rules proposed by BNSF in its original petition letter.
11. FRA may modify or rescind this waiver based on information regarding safety of rail operations or for noncompliance with any conditions.
12. This pilot test waiver lasts 2 years, with an option to extend based on test results, and BNSF applies.

Federal Railroad Administration

11

FRA-2012-0091

- The text of FRA's Decision Letter, outlining these conditions may be found in the public docket
- Go to www.regulations.gov and search for Docket Number FRA-2012-0091

Federal Railroad Administration

12

Test Committee

- Comprised of about 24 participants
  - ➢Class I RR's (CPR, CNR, UP, BNSF), FRA, AAR, GE, Wabtec, NYAB and union representation
- Operate under an approved FRA Pilot Test Waiver
- Conform to an approved committee Test Waiver Plan
- Data is collected on submitted trains and provided in a report to the committee on a monthly basis
- Committee convenes at a minimum every six months

Federal Railroad Administration

13

Test Committee

June 27, 2014 Test Committee Meeting

Members reviewed and commented on Test Train summary for first season of test

- Agreed to vote electronically on all further business. Following the meeting committee voted to accept:
  - ➢ Meeting minutes of 6/27 meeting
  - ➢ Data set summary for first 22 candidate trains
  - ➢ Committee to open up waiver to other roads for submission of candidate trains and data collection
  - ➢ Continue 90 CFM Waiver for another 6 month period

Federal Railroad Administration

14

Opening of Test Waiver to Other Railroads

- Canadian Pacific requested consideration to operate 90 cfm trains in the United States (FRA-2013-0040)
- FRA denied the petition as FRA felt it could not monitor two test waivers simultaneously. CPR was encouraged to continue its participation in BNSF's test. Pending test results, CPR could reapply at a later time
- Following the committee's report of the first season results, and its request for more candidate trains; CPR resubmitted its request

Federal Railroad Administration

15

Opening of Test Waiver to Other Railroads

- Canadian Pacific's request was accepted by FRA at the September meeting of its Railroad Safety Board.
- Test committee to discuss how the CPR is to implement testing in the US, including meeting the conditions of the BNSF waiver and the current approved test plan.
- This past week Canadian National filed its own request to join the test waiver. CNR's request will be considered by the Railroad Safety Board at its October meeting

Federal Railroad Administration

16

Opening of Test Waiver to Other Railroads

regulations.gov
Your Voice in Federal Decision-Making

- Copies of these requests are posted to the public docket at www.regulations.gov
- Interested parties may review all posted items related to the docket by entering the docket number (FRA-2012-0091) into the search engine and downloading

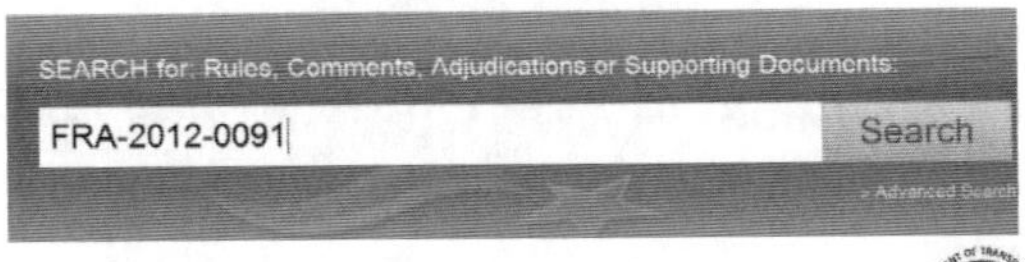

Federal Railroad Administration

17

Timeline

HOW HAVE WE PROGRESSED?

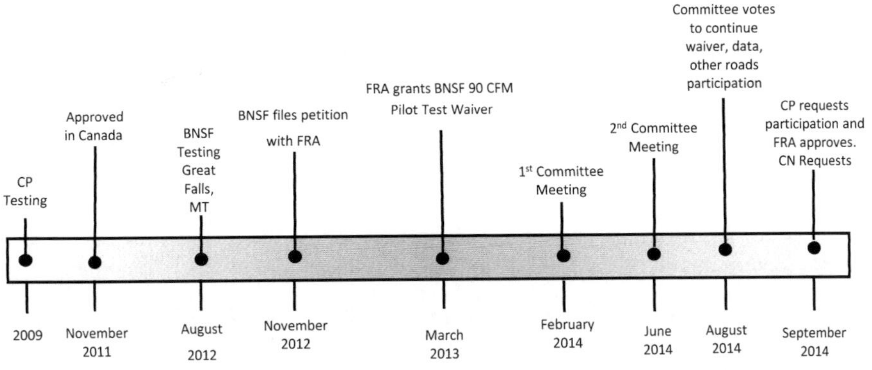

BNSF RAILWAY

18

Pilot Test Waiver Plan- Instructions

- Specific Subdivision General Orders issued on Powder River, Twin Cities and Montana Divisions with rules outlining requirements of waiver.
- Instructions to include any DP train operating above 60 CFM to notify respective RFE of any such occurrence(s) at completion of trip and before tying up.
- RFE secures downloads (both lead and controlling remote(s)), reviews and forwards to Superintendent of Operating Practices (SOP) and Director of Train Handling, Operating Practices (DTH) in Ft. Worth to review train operations during period(s) of greater than 60 CFM.

Pilot Test Waiver Plan- Instructions

- Review will include the following items though will not limited to:
  - ❖ Train length,
  - ❖ Duration/distance of CFM greater than 60 CFM
  - ❖ All air brake usage while greater than 60 CFM
  - ❖ Variations in "Stable Released" Air Flow relative to Time Of Day, Location, Speed, and Temperature
- SOP and/or DTH documents any exceptions taken to the above metrics
- DTH verifies ambient temperatures via 'Weather Underground' website across location where train operated greater than 60 CFM.
- Owning SOP tracks/documents 'excepted' trains for their respective Division - forwarding monthly to DTH.

Pilot Test Waiver Plan- Instructions

- If a train encounters braking issues while operating within the waiver parameters of 60 to 90 CFM, owning SOP will ensure to have train inspected and provide results to the DTH.
- DTH to file monthly report to test committee representatives for feedback on analysis.
- Engineer Standard Interview questionnaire suggestions (modeled after CPRR):
  - ❖ Overall Brake System Performance: 1 □ 2 □ 3 □ 4 □ 5 □ 6 □ (1 is extremely poor, 6 is excellent)
  - ❖ Overall Recharging Performance: 1 □ 2 □ 3 □ 4 □ 5 □ 6 □ (1 is extremely poor, 6 is excellent)
  - ❖ Did You Experience Any Sticking Brakes: Yes □, No □ if yes, (stn, sub/mile)
  - ❖ Did You Experience Any Kickers (UDEs): Yes □, No □ if yes, (stn, sub/mile)
  - ❖ Did You Experience Any Unintentional Releases: Yes □, No □ (stn, sub/mile)
- Other Comments/Concerns (to include any detailed information regarding answers above)

BNSF RAILWAY

6/5/2015 21

Pilot Test Waiver Plan- Operating Rules

- When a DP train's brake pipe exceeds a combined air flow rate of 60 CFM, the DP remote consists' brake valve must remain 'CUT-IN' at all times. If for any reason a DP remote consists' brake valve is 'CUT-OUT', the train must stop and be inspected for leaks within the train line. Moreover, the train may not proceed until brake pipe air flow rate reaches 60 CFM or lower (the same as conventional train requirements).
- When a DP train's brake pipe exceeds a combined air flow rate greater than 90 CFM for the entire train and/or more than 60 CFM for any one locomotive consist in the train when an automatic brake handle is in the release position, other than during an intended brake application and/or brake release activity, then the train must stop and be inspected for leaks. Additionally, the train may not proceed until the combined air flow rate reaches 90 CFM or lower.
- The brake pipe pressures must be within 15 pounds per square inch (psi) of standard operating pressure when the brakes are released and fully charged.

BNSF RAILWAY

6/5/2015 22

Pilot Test Waiver Plan- Operating Rules

- If the combined brake pipe air flow is greater than 60 CFM and any DP remote consist fails en route and loses air compressors (source of air), then the train must be stopped and may not proceed until the air compressor is repaired, a source of air is regained, or the air flow reaches 60 CFM or lower.
- DP trains with a combined air flow rate greater than 60 CFM will not be allowed to operate over any grades greater than 1.5 percent that exceeds 3 miles (descending and/or ascending).
- Air flow meters on both LEAD and DP REMOTE must be within 92 day calibration requirement.

BNSF RAILWAY

6/5/2015 23

Data Overview

Train Size	Min	Max	Avg
Tonnage	2,327	16,882	10,033
Length (ft)	5,083	7,024	6,415

Temp	Min	Max	Avg
Ambient (F)	-35	-2	-18

Total Trains	22
Exceptions	0
UDE's	0
Unintentional Releases	0
Hard Braking/Over-reductions	0

22 DP Candidate Trains

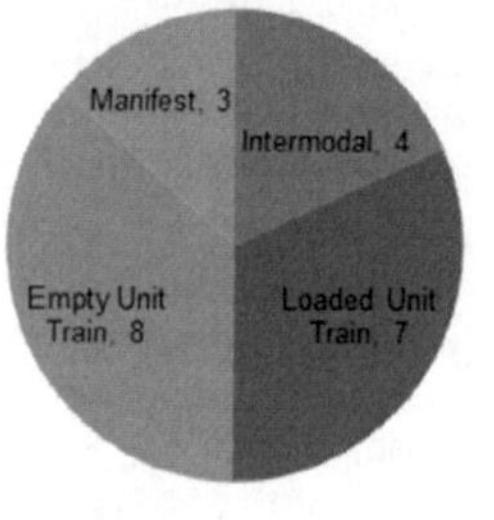

BNSF RAILWAY

6/5/2015 24

Results

- As temperature decreases flow increases. Specifically seen at temperatures below 0 degrees.
- No difference in BP performance across various train types.
- Brake applications and releases ranging from a minimum reduction to a full service application were made without any problems at brake pipe leakage rates of 60 CFM and 90 CFM
- No significant differences were noted between the brake applications made at 90 CFM versus the brake applications made at 60 CFM or in between.
- Several trains did exceed the 90CFM but were secured till flow resided.

BNSF RAILWAY

6/5/2015 25

Next Steps

- Committee approved in August meeting to continue waiver for another 6 month period
- Approval by committee to open up current waiver to other railroads for submission of candidate trains and data observation.
- Participation of CPR this winter and in addition acceptance of CNR into the program.
- 2014-2015: Increase the number of candidate trains and data analyzed.
- Proposal for future amendment to regulation.

BNSF RAILWAY

6/5/2015 26

Questions?
Federal Railroad Administration
BNSF RAILWAY
27

Mark Kraeling, GETS

Mark Kraeling has over 20 years of experience in the embedded systems industry, primarily focused on embedded real-time, communications, and safety critical systems for the automotive and transportation industries. He is a regular speaker at embedded software conferences worldwide, and has published various papers and articles focused on embedded systems development over the course of his career. Mark co-edited and authored a book titled Software Engineering for Embedded Systems: Methods, Practical Techniques, and Applications, published in April, 2013. He has BSEE, MSEE, and MBA degrees from Rose-Hulman, Arizona State, and Johns Hopkins Universities, respectively. Mark also works with various subcommittees in industry organizations such as the Association of American Railroads to develop architecture and communication-based standards. Mark started his career at Cummins, Inc. developing real-time engine controls. He is currently a product manager and system architect at GE Transportation in Melbourne, Florida.

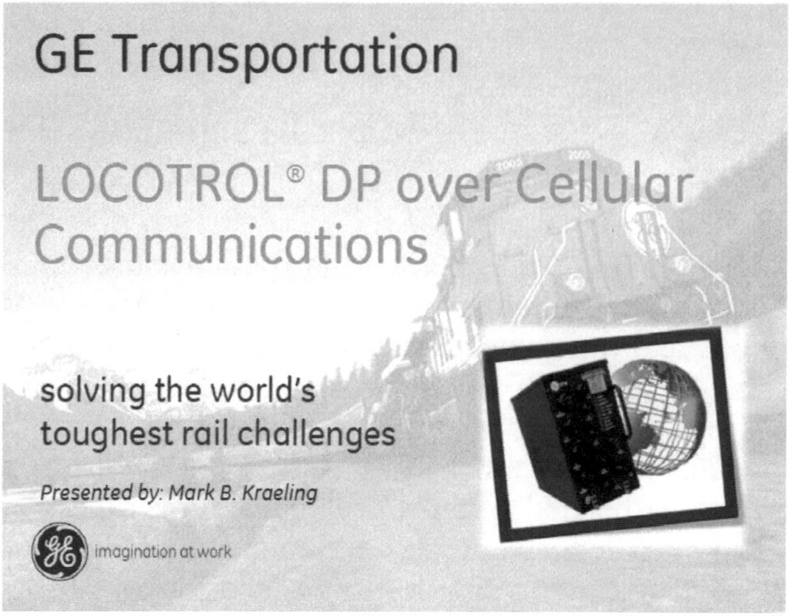

Speaker Background

Mark Kraeling

- GE Transportation Cab Electronics
  - Products for Locomotive Onboard
  - Based in Melbourne, FL
- Architect for Network Solutions
  - Wired/Wireless Communications
  - Linux and ARM based Designs
  - Virtualization Solutions

imagination at work

2
RSI/CMA Canadian Rail Summit 2014
9/22/2014

Some of My Activities in 2014...

Mark Kraeling

- AAR-9154a Intra Consist Routing
- AAR TAG Demonstrations
  - Upload/Download Box
  - Consist Communication
- AAR S-9101 ACC Platform
  - Technical Architect
  - New Cards / Configurations

imagination at work

3
RSI/CMA Canadian Rail Summit 2014
9/22/2014

Agenda

- DP Overview
- 450MHz Radio Diversity Option
- Cellular Communications
  - Testing / Evaluation
  - Phased Approach
  - Production Implementation

imagination at work

4
RSI/CMA Canadian Rail Summit 2014
9/22/2014

DP Overview

imagination at work

LOCOTROL® Distributed Power

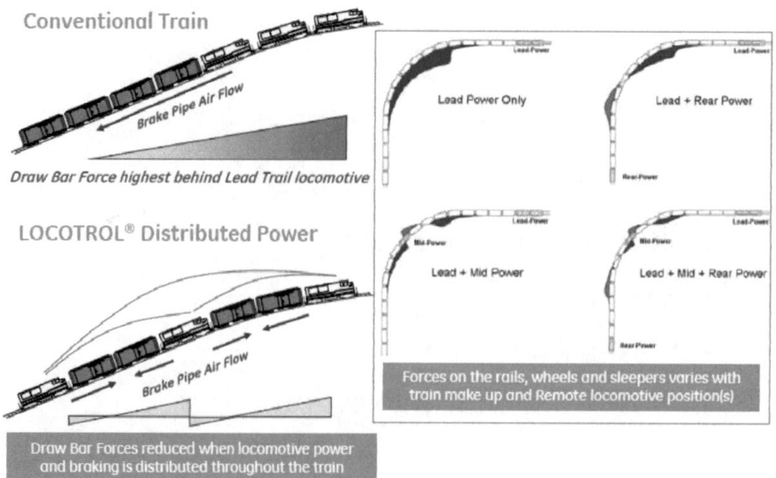

imagination at work

6
RSI/CMA Canadian Rail Summit 2014
9/22/2014

LOCOTROL® Distributed Power Worldwide

BNSF	5,246
	3,588
Canadian Pacific Railway	787
	532
CN	525
NS	503
CSX	475
Brazil	1918
China	960
Mexico	466
Australia	270
Other	680
	15,426

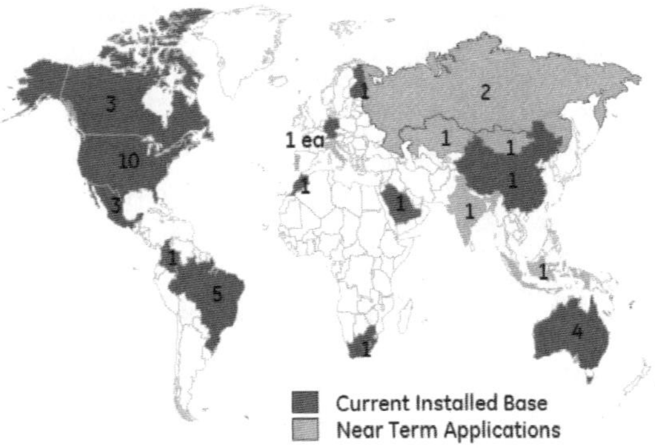

imagination at work

450MHz Radio Diversity Option

imagination at work

450MHz Radio Diversity Testing

- Colton, CA to Bakersfield, CA
- 18+ Hours of Running Time
- Trailing unit behind lead with hardware modification for "true" analog diversity

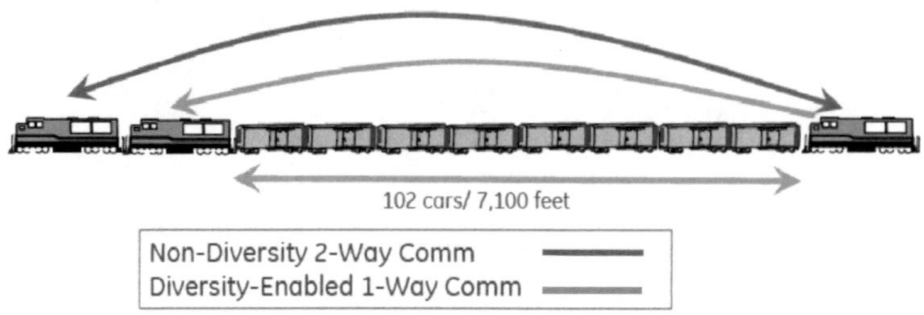

imagination at work

450MHz Radio Diversity Types

- Most DP Systems, Dithering Antennas

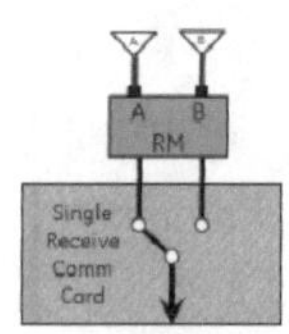

- New DP Dual Receive Card, using analog signals

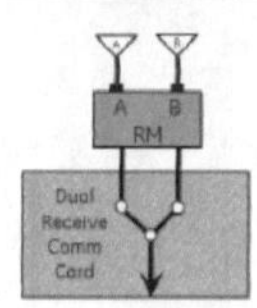

- LOCOTROL™ XA – true digital diversity

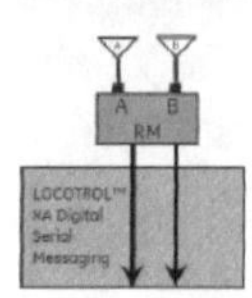

imagination at work

Cellular Communications

Testing/Evaluation

imagination at work

Testing / Evaluation

- Use similar type of run (hills/etc) as diversity testing
- Radio 450MHz communications will be primary path
- LOCOTROL™ DP systems will simultaneously communicate over cellular – Track Messaging

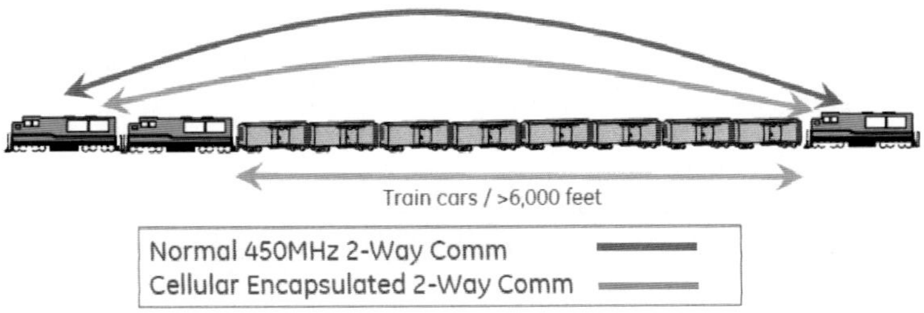

imagination at work

Testing / Evaluation

- Units DP radio link as usual
- Cellular link created after radio link – using DP data to link together
- Messages sent/received over both links will be logged for data analysis
- Could extend to test two cellular paths at the same time (such as AT&T and Verizon)

imagination at work

13
RSI/CMA Canadian Rail Summit 2014
9/22/2014

Testing / Evaluation Results

- X-Axis will be time, marked with mile marker
- Y-Axis will show messages sent/received by carrier, as a percentage

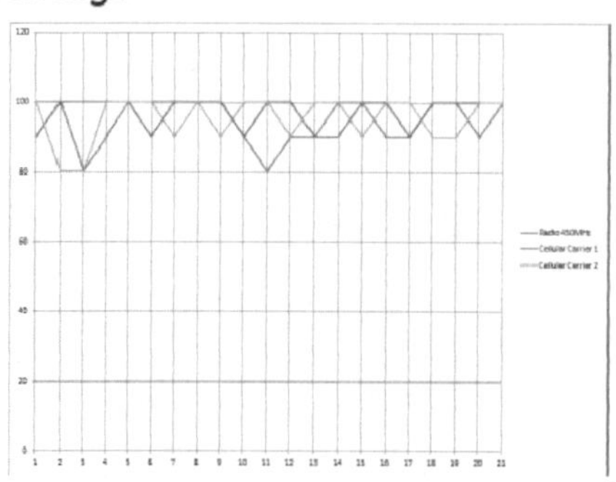

imagination at work

14
RSI/CMA Canadian Rail Summit 2014
9/22/2014

Cellular Communication

Phased Approach

imagination at work

Phase 1

- Connect Ethernet of LOCOTROL™ XA to Locomotive Network
- Configure GoLINC ACC to route correct port traffic to other LOCOTROL™ XA unit in remote consist over cell

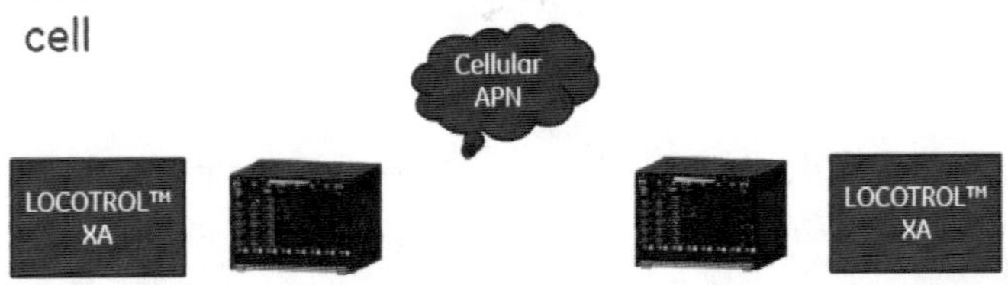

imagination at work

Phase 1.5

- Configure GoLINC ACC to route correct port traffic to other LOCOTROL™ XA unit in remote consist over two cellular paths at the same time

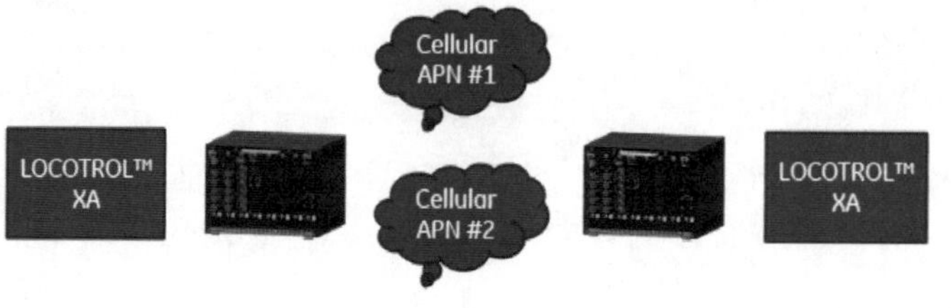

imagination at work

Phase 2

- Using same networked connection, LOCOTROL™ XA uses local ITCM messaging stack

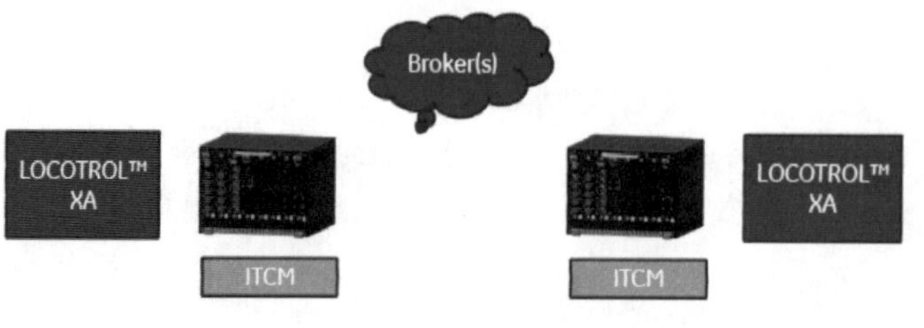

imagination at work

Cellular Communications

Production Implementation

imagination at work

Production: Diagram

- Units will still link using 450MHz DP radios
- ITCM messaging will be activated periodically during good communications, and continuously during bad

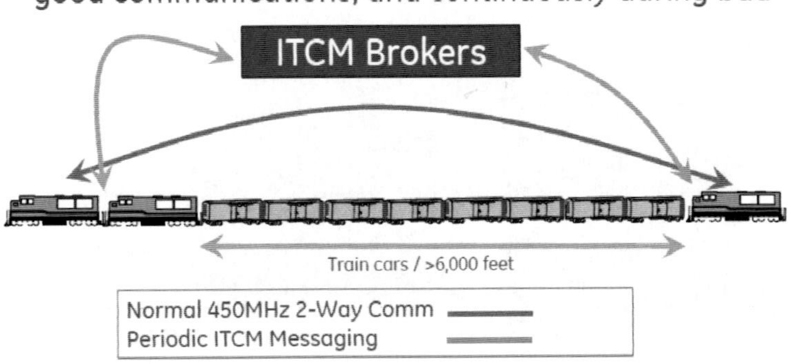

imagination at work

EMP Messaging

- LOCOTROL™ system will use EMP Class D messaging
- Discussions have already begun with MeteorComm to assign a message ID for the system, likely will be "DP"

EMP Message Header	New DP Messaging	Existing DP Message	New DP Footer	EMP Message Footer

imagination at work

ITCM Messaging Diagram

- LOCOTROL™ messages will be delivered over any of the available ITCM messaging paths, such as WiFi or cell

imagination at work

Conclusions

- Improving communications for LOCOTROL™ DP will be a key driver for the rest of 2014 and 2015
- Testing will show percentage of improvement using cellular communications in addition to 450MHz communications
- Production release will use customer preferred network communications (such as ITCM) in addition to 450MHz communications path

imagination at work

Q/A

LOCOTROL™ North American Product Manager

Brian Schroeck

Brian.Schroeck@ge.com

Cab Electronics Onboard Architect

Mark Kraeling

Mark.Kraeling@ge.com

imagination at work

imagination at work

Gary Wolf,
Wolf Railway Consulting

Gary Wolf received a BS in Electrical Engineering from Ohio University and an MS from Georgia Tech. He began his railroad career in 1970 in the Mechanical Engineering Department with Southern Railway. In 1973 he moved to the Industrial Engineering group. In 1976 he was appointed to a new position as track/train dynamics engineer in Southern's Operation Research Department. At this time he became responsible for the study and analysis of train derailment and accident investigations, track/ train dynamic studies, new locomotive and car technologies, and rail capacity studies. In 1986 he was promoted to Manager of Operations Research at Norfolk Southern. In 1987 Gary left Norfolk Southern and founded Rail Sciences, Inc. Over the next 23 years Rail Sciences expanded to over 20 employees with two locations and a global practice. In 2010 Rail Sciences was acquired by TUV-Rheinland, a German based Engineering and Testing firm. In 2013 Gary left TUV and founded Wolf Railway Consulting, practicing in the area of train, track and vehicle dynamics.

Gary has spent nearly 39 years of his railroad career on the study and determination of train accident causes, especially derailments. In addition, Gary has been active in the field of vehicle dynamics and wheel/ rail interaction and vehicle dynamics. He is a member of the ASME Rail Transport Division General Committee, AREMA, The Air Brake Association, and was elected President of the International Railway Operating Officers Association for 2005-06. In addition, Gary served on numerous AAR committees during the Track/Train Dynamics research program of the 70's and 80's.

Gary was president of the IAROO in 2006, has been a member of the executive committee for over 20 years, and has attended nearly every IAROO fall conference since the late 1970's. In addition, Gary has presented 25 technical papers at the IAROO annual meeting since 1980. In 1985 Gary presented a paper dealing with the analysis and prevention of low speed derailments, and his presentation today will be an update on that topic.

Prevention of Yard, Terminal & Industry Derailments

Presented to:

IAROO Montreal 2014

Presented by:

Gary Wolf

Wolf Railway Consulting
2838 Washington Street
Avondale Estates, Georgia 30002
404-600-2300
www.wolfrailway.com

WOLF Railway Consulting

1

2013 FRA Reportable Derailments

Mainline Derailments 389

Yard & Industry Derailments 891

WOLF Railway Consulting

2013 FRA Derailments

- $ 186 M for Mainline Derailments
- $ 89 M for Yard & Industry Derailments

A Huge Opportunity for Cost Savings!

3

2013 FRA Main Track Derailments

- 389 Main Track Derailments
  - 116 Derailments at speeds between 1 – 9 MPH

4

Yard & Terminal Derailments Subject Outline

- Human Factor Causes
- Track Causes
- Mechanical Causes

WOLF

Specific Human Factor Derailments

- By-Passed Couplers
- Blind Shoves
- Switch & Switch Point derailments
- Run Through Switches

WOLF

Switching Operations Fatality Analysis (SOFA)

- FRA,AAR, UTU, BLE, ASL&RRA
- 76 Fatal Switching accidents investigated
- Jan 92 - July 98 - Detailed Analysis of all available factors (Weather, Site, Personal info, Visibility, Communication issues, PPE, D&A tests, Fatigue, Time & date)
- Also limited data from 1975-1991

Interesting Facts SOFA Study

- Most Fatals occurred between 0-6 hours worked; few between 7-12 hours
- 74% of fatals occurred with 3 man crew;only 15% with 1 and 2 man crews
- Most fatals in Dec,Jan, June, July
- Most fatals on Tuesday, weekends low
- 36% 1st shift, 36% 2nd shift, 26% 3rd
- No real trend by hour of day

Interesting Facts SOFA Study

- Operating rule violations ranked as highest contributory cause factor
- While the SOFA study did not specifically address derailments, many of the human failure actions can also lead to derailments.

WOLF
Railway Consulting

As a result of an increase of human factor derailments and employee injuries in the early 2000's, FRA promulgated switching rules contained in sections 217 and 218 of the federal code

WOLF
Railway Consulting

10

CFR 218 – Railroad Operating Practices

- Part 218 of the Code of Federal Regulations specifies the minimum requirements for railroad operating rules and practices.
- This Part was modified on February 13, 2008 (and additionally updated on June 16, 2008) to add a new Subpart F - Handling Equipment, Switches and Fixed Derails.
- The detailed requirements are described in Sections 218.99, 218.101, 218.103, 218.105, 218.107 and 218.109.

WOLF Railway Consulting

It is clear from the language of FRA 218 that Job Briefings are an essential part of safe switching procedures.

WOLF Railway Consulting

12

Job Briefings Essential for Switching Safety & Efficiency

- Plan the job briefing
- Conduct the job briefing
- Special conditions
  - Complex Jobs
  - changes in conditions (e.g., weather)
- Follow-up by supervisors; Efficiency Testing

WOLF Railway Consulting

Planning Job Briefing

- Develop Work Plan
  - Know Location & Conditions
  - Break tasks into step- by-step
  - Determine tools/equipment needed
  - Determine safety rules applicable
- Determine Potential & Existing Hazards
  - Weather; Visibility; Traffic Conditions
  - Conflicts with plant traffic (off-road equip)
  - PPE required
  - Time of day; time of year
  - Tracks to be used

WOLF Railway Consulting

Conduct Job Briefing

- Explain to all employees
  - Who, What, Where, When, Why, How
- Discuss potential hazards
- Make Work Assignments
- Explain special tools or methods
- Insure Understanding of all crew members

Special Conditions

- Complex Jobs
  - Do one part at time
  - More briefings as job progresses
- Changes in Conditions
  - Weather
  - Traffic Conditions
  - Time of day
  - Customer Requests

Supervisory Follow-up

- Check to see that plans are being followed
- Check for rules compliance
- Look for other hidden hazards
- Make sure personnel capable of carrying out assignments
- Efficiency Testing

Prevention of Human Factor Derailments

(Note: the following recommendations and guidelines are not meant to supplant any individual railroad rules or operating practices)

18

Avoiding By-Passed Coupler Derailments

- Personnel Always Perform Job Briefing
  - Discuss Coupling & Uncoupling in Curves
  - Insure New Employees Understand Facility
- Identify Cars Most Susceptible to Passed Couplers (Long Couplers; centerbeams, pig flats, autoracks)
- When Shoving into Track, Attempt to Place Long Cars Well onto Straight (Tangent) Track Before Uncoupling

WOLF

Avoiding Passed Coupler Derailments Continued

- If Switching (Kicking) on Lead, Insure Cars Uncoupled on Straight Track
  - If Lead Track has Tangent and Curved Portion, Identity Where Desirable to Uncouple
    - Example - Paint White Line at Start of Curve (Do Not Uncouple Beyond White Line)

WOLF

Avoiding Passed Coupler Derailments Continued

- Check for Open Knuckles (Insure both not closed)
  - If Must Open Knuckle:
    - If cannot open knuckle with cut-lever, Insure Equipment Separated by at Least One Half Car length (25') and All Movement Stopped Before Opening Knuckle
    - Stand to Side Rather Than in Front of the Coupler With One Foot Outside the Rail
- Whenever Possible, Leave Knuckles Open For Subsequent Switching Moves

WOLF

Avoiding Passed Coupler Derailments Continued

- Check Drawbar Alignment When Coupling
  - If Drawbars Misaligned:
    - Insure Equipment Separated by at Least One Car length (50') and All Movement Stopped
    - Three Step Protection – Red Zone
    - Insure All Personnel (Including Other Jobs in Area) Know Drawbar(s) Are Being Straightened and Crewmember Between Cars
    - Get Assistance or Tool if Difficult to Move (Example - Drawbar Strap)

WOLF

Avoiding Passed Coupler Derailments Continued

- Stretch Cars Before Making Shoves
  - Look for Separations Indicating Possible Passed Couplers
  - If Found, Move to Location to Insure Coupling Properly Made
- Visually Monitor Couplings Whenever Possible
- Avoid Blind Shoves

Avoiding Blind Shove Derailments

- Whenever Possible and Appropriate Ride or Walk Along Side of Leading Car
- When Shoving Clear Track Remain at Location Where Lead Car is Continuously Observed Until Stop
- Always Check Switches and Derails on a Track being Shoved (do not assume properly lined)

Avoiding Blind Shove Derailments Continued

- Maintain Constant Communication With Engineer During Entire Shove
  - If using Radio, Movement Must be Stopped in Half the Distance if Communication Not Received
- If a Non-Crewmember is Protecting the Shove by Radio, Lead Car Movement Should be Continuously Observed by Crewmember or Non-Crewmember

Avoiding Blind Shove Derailments Continued

- When Engines (Jobs) are Working at Both Ends of a Track or Tracks, Crews Switching Must have a Clear Understanding of Movements to Be Made (General Switching Instruction)
  - Crews Should not Shove Toward Another Job Working the Other End of a Track or Tracks Without Crewmember Riding on or Walking beside Lead Car of Movement

Avoiding Blind Shove Derailments Continued

- When Shoving Car(s) into a Spur Track
  - Stop Movement 150' From End of Track (safety stop)
  - Apply Hand Brake(s) When Necessary To Control Slack
  - Crew Member Should Precede any Further Movement, and Continuously Control Movement (Example, by Radio)

Avoiding Switch and Switch Point Derailments

- Personnel Handling Switches Should Perform the Following:
  - Insure Switches and Derails are Properly Lined for Intended Route
  - When Lining Switches, Inspect Points to Insure that Points Fit Properly and Target (If Equipped) Corresponds With Switch Position

Avoiding Switch and Switch Point Derailments

- Whenever Possible, Remain at Least 20' From Switch Stand While Movement is Closely Approaching and Proceeding Over Switch (May be Exception If at Switch When Kicking or Dropping Cars)
- Insure that All Wheels of Car have Sufficiently Cleared the Switch Points Prior to Lining Switch

Avoiding Switch and Switch Point Derailments

- If Permitted to Drop Cars, Prior to Starting Movement Line and Secure Switch
- If Switch Equipped with Lock, Lock Switch & Test Lock When Movement(s) Completed
- If Switch Equipped with Hook, Place Hook in Hasp When Movements Completed
- Whenever Possible, Personnel on Engine Should Verify that Switches and Derails Near Engine are Properly Lined

Switch Point Protectors

- Recommended on Yard Switches With Standard Switch Point (Not Needed with Samson Point if Used in Yard Switches)
- Protects Point Used for Turnout or Curved Side (Switch Reverse) Movement

WOLF

Switch Point Protectors Continued

- Priority for Switches Selected
  - Switches with High Use in Reverse Position
  - Highly Used Lead Switches
  - Speed Must Be 20 MPH or Less
- Suggest Protector With Adjustable Shoe
  - Allows for Adjustment with Wear

WOLF

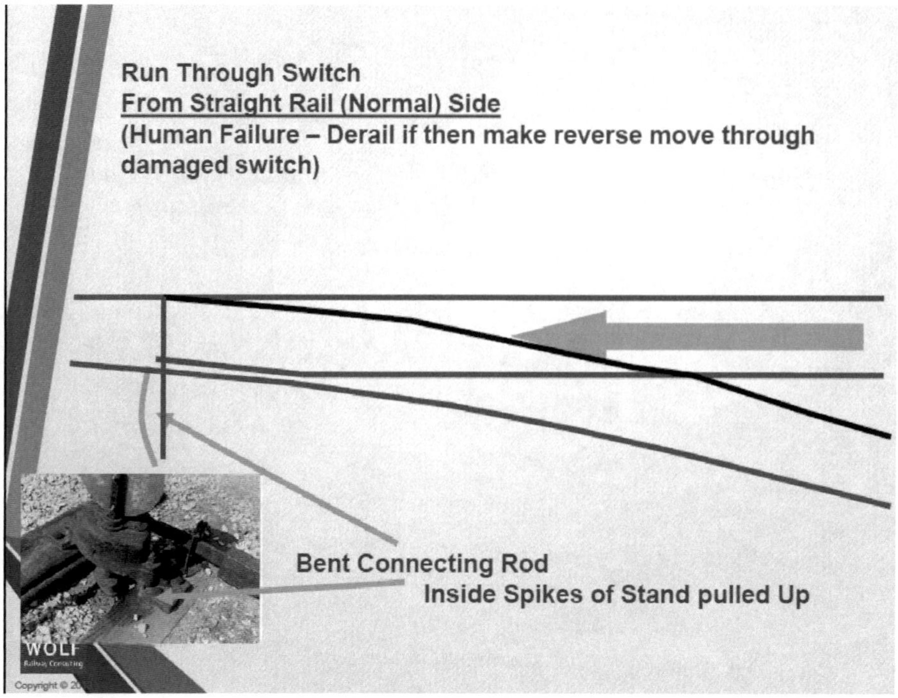
Run Through Switch
From Straight Rail (Normal) Side
(Human Failure – Derail if then make reverse move through damaged switch)
Bent Connecting Rod
Inside Spikes of Stand pulled Up
WOLF

Bent Connecting Rod
WOLF

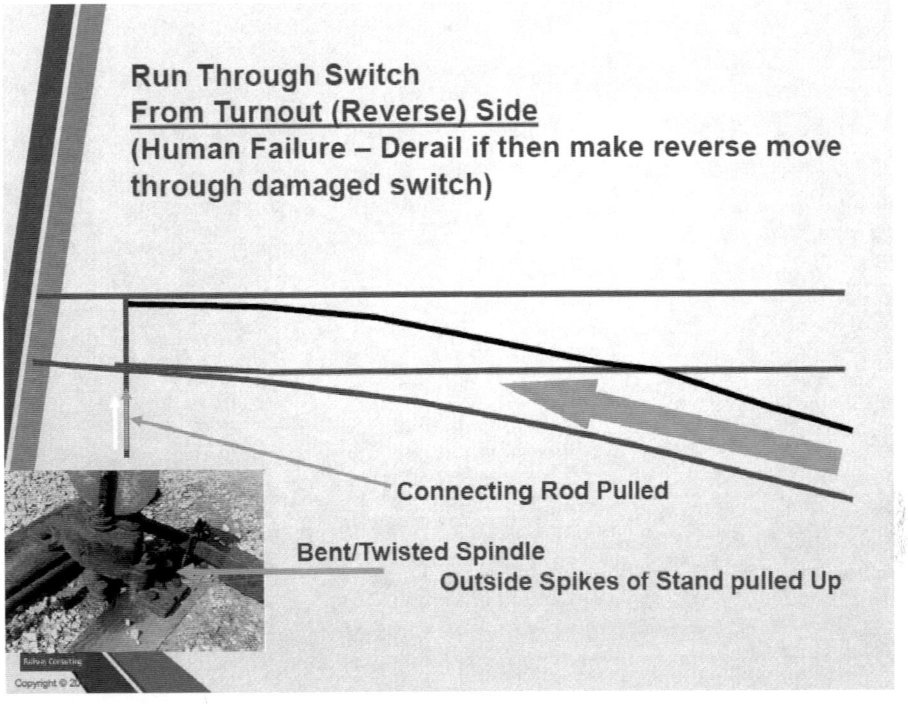
Run Through Switch
From Turnout (Reverse) Side
(Human Failure – Derail if then make reverse move through damaged switch)
Connecting Rod Pulled
Bent/Twisted Spindle
Outside Spikes of Stand pulled Up

Note Twist on spindle shank
WOLF

Twisted Spindle
WOLF

Approximate amount of
Angular twist on spindle
WOLF

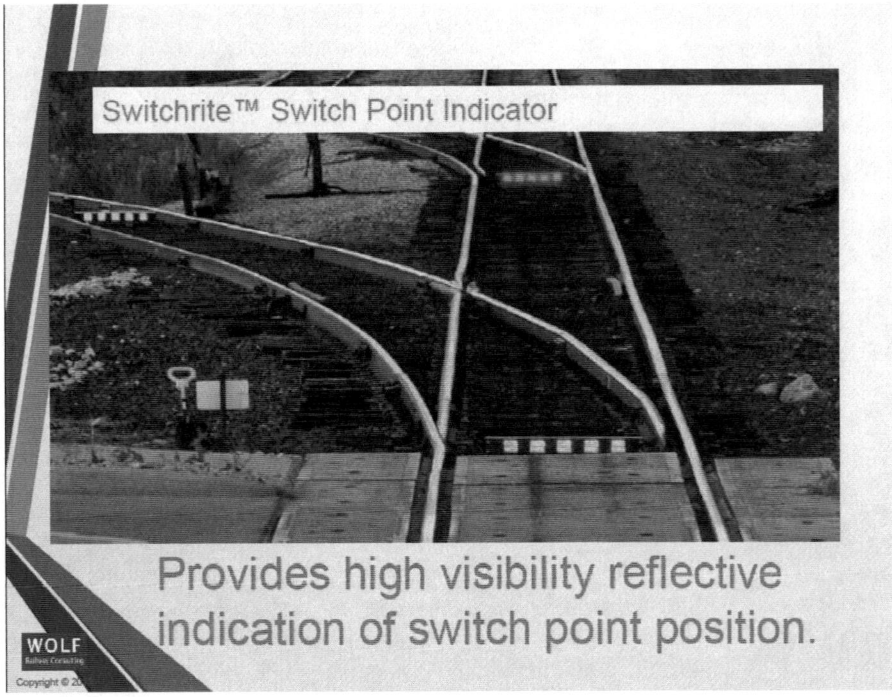
Switchrite™ Switch Point Indicator
Provides high visibility reflective indication of switch point position.
WOLF

Run through Switch Indication
WOLF

Aldon Company Inc.
Switch Cube™ Indicator
WOLF

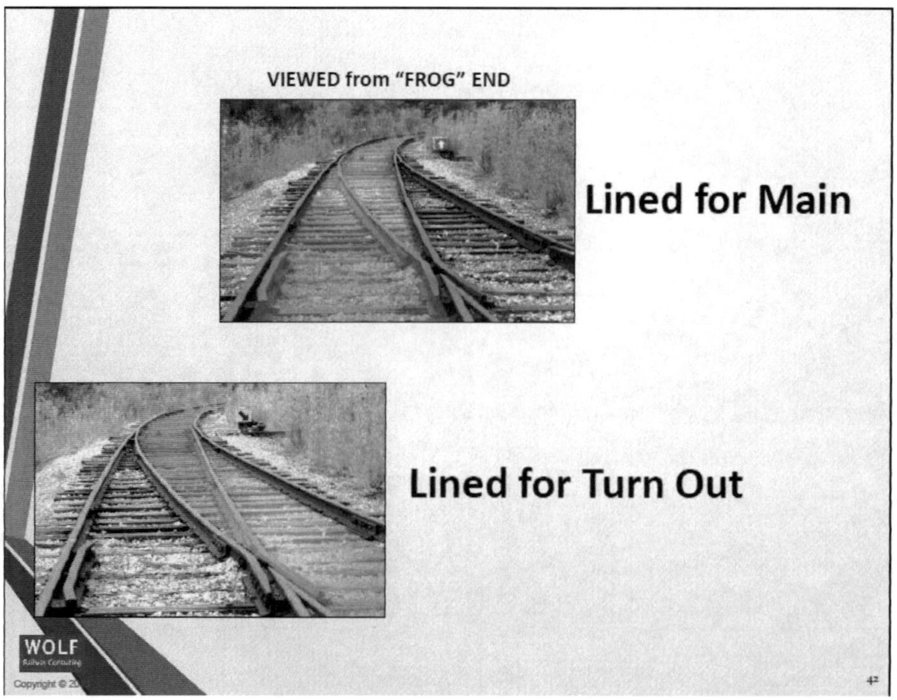
VIEWED from “FROG” END
Lined for Main
Lined for Turn Out
WOLF

Track Issues

- Wide Gage Derailments
- Facing Point Switch Derailments
- Track Geometry

Why focus on Wide Gage Derailments?

- Most used cause code in the FRA Accident Database
- ~20% of all track caused derailments attributed to T110, T111 Wide Gage
- Many derailments listed as "wide gage" incorrectly blamed on "wide gage".
- Other Causes such as mechanical and train handling often not listed.

WOLF
Railway Consulting

"Wide Gage description possibly the least understood of all derailment causes"

45

WOLF
Railway Consulting

"The true root cause(s) of wide gage are generally not reported. Not all 'wide gage' derailments are track caused. Many 'wide gage' derailments have mechanical and train operations root causes"

46

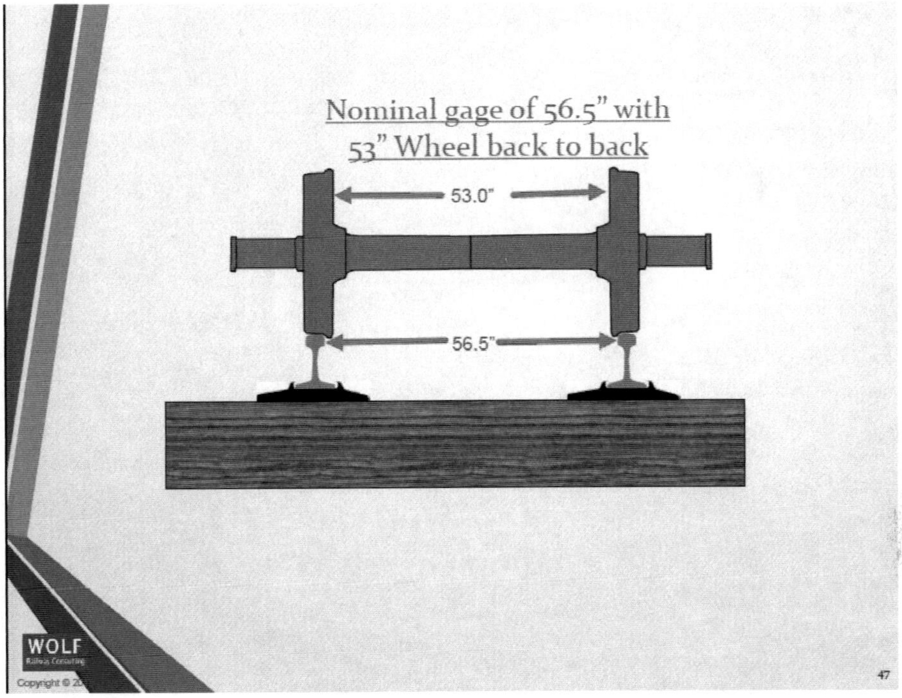
Nominal gage of 56.5" with
53" Wheel back to back
53.0"
56.5"
WOLF
47

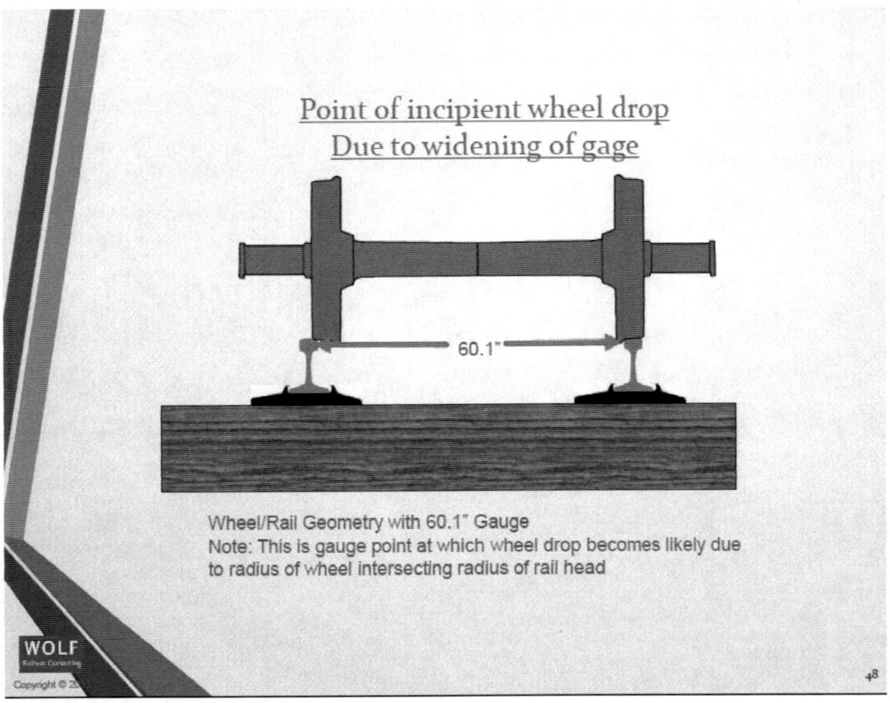
Point of incipient wheel drop
Due to widening of gage
60.1"
Wheel/Rail Geometry with 60.1" Gauge
Note: This is gauge point at which wheel drop becomes likely due
to radius of wheel intersecting radius of rail head
WOLF
48

Wide Gage due to lateral
Translation of rail
WOLF
49

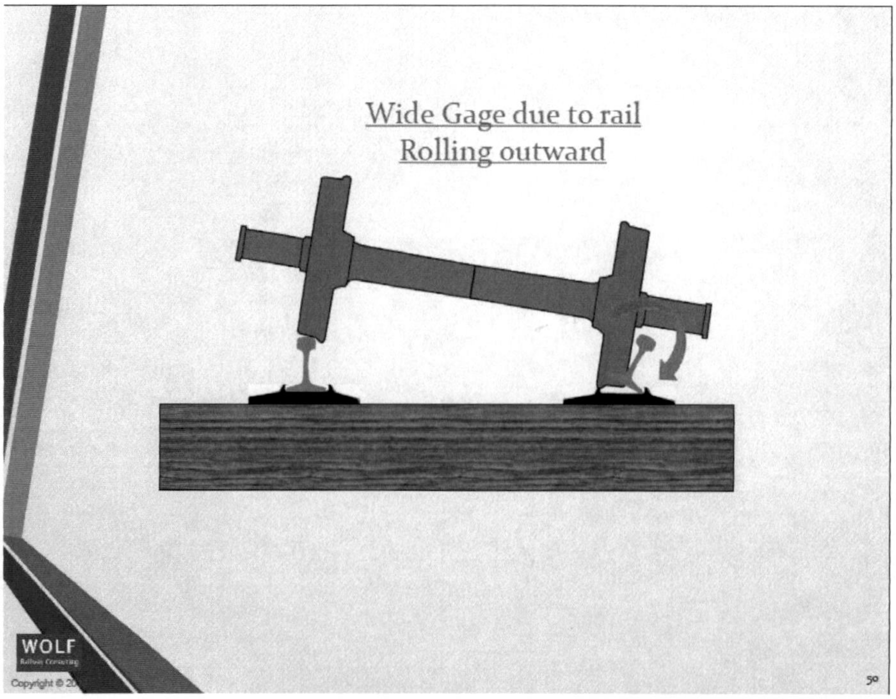

Wide Gage due to rail
Rolling outward
WOLF
50

Wide track gage is caused by:

- Lateral translation of one or both rails, due to:
  - Deformation of wood fibers holding spikes; elongated spike holes; spike kill
  - Broken screw spikes or cut spike fasteners
  - Worn shoulders on tie plates
  - Gage face wear on the rail
- Rail rolling outwards, due to:
  - Differential tie plate cutting; rail cant
  - Loose fasteners; weak spiking; insufficient spiking
  - Poor wheel contact geometry toward field side of rail head
  - Hollow worn wheels contacting field side of rail head
- A combination of lateral translation and rail roll

WOLF Railway Consulting

51

How does Gage Widen?

- Progressive failure (fatigue) over weeks or months
  - *"Death by a thousand small cuts..."*
    - ✓ Spike kill/ tie deformation over time
    - ✓ Fastener failure
    - ✓ Differential plate cutting; rail cant
    - ✓ Geometry defect
- Singular catastrophic event
  - *"Death by a single blow to the head..."*
    - ✓ Bypassed couplers during switching
    - ✓ Excessive buff (compressive) force; slack action
    - ✓ Use of non-aligning couplers; locomotives
    - ✓ Truck warp; locked bolster on truck

WOLF Railway Consulting

52

What causes gage to widen?

- **Excessive lateral pressure against the rail due to:**
  - Incorrect curve elevation
    - ✓ Insufficient elevation causing pressure on high rail
    - ✓ Excessive elevation causing pressure on low rail
  - Pre-existing wide gage allowing greater wheelset angle of attach
  - Horizontal alignment kinks causing increase in flanging force
  - Poor steering due to poor wheel – rail contact geometry

WOLF Railway Consulting

What causes gage to widen (con't)?

- **Excessive lateral pressure against the rail due to:**
  - Excessive compressive buff force in train
    - ✓ Excessive dynamic braking
    - ✓ Excessive independent braking
    - ✓ Excessive shoving amperage
    - ✓ Bypassed couplers
  - Stiff truck condition; excessive turning moment locking up bolster rotation
    - ✓ Centerbowl not lubricated
    - ✓ Centerbowl rim contact or fouling
    - ✓ Tight side bearings
    - ✓ Improper centerplate repairs
  - Warped truck condition
    - ✓ Lack of truck squaring force due to wedge rise; bolster pocket wear; column plate wear

WOLF Railway Consulting

Let's review how our maintenance strategies and operating procedures can contribute to increased lateral pressure against both the high and low rail of curves...

WOLF Railway Consulting

55

Superelevation and adverse lateral forces

- **Superelevation is added to curves to allow trains to operate at normal track speed without placing abnormally high lateral and vertical forces against the high rail.**
- **Insufficient elevation can lead to excessive lateral pressure against the high rail in curves**
- **Excessive elevation places adverse vertical forces on the low rail of curves. Excessive vertical forces on the low rail of curves also increases the development of large lateral creep forces toward the field side.**

WOLF Railway Consulting

56

Pre-Existing Wide gage

WOLF
Railway Consulting

57

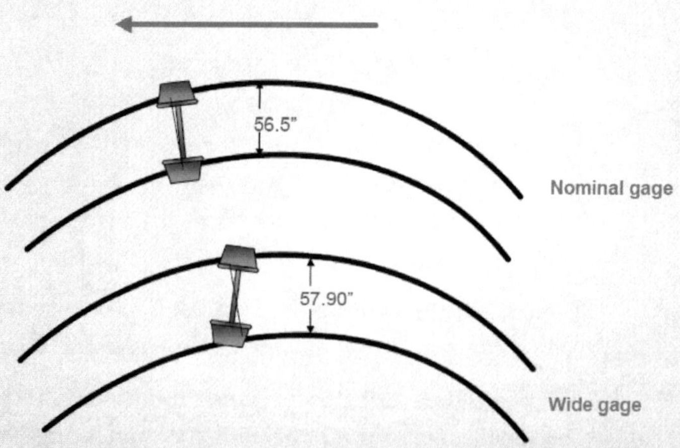

Pre-existing Wide Gage can allow greater wheelset angles of attack, especially for hollow worn wheels;
Greater angles of attack can cause greater lateral creep forces.
Wide gage also can lead to poor wheelset steering and adverse contact geometry

WOLF
Railway Consulting

58

Curve Alignment Kinks/Irregularities

WOLF
Railway Consulting

59

Plug Rail on High Rail in Middle of Curve

WOLF
Railway Consulting

60

POD
Note change in alignment in vicinity of POD. Measurements show curve increases to nearly 15 degrees at this location
Looking from near station 3 toward the POD (Bag laying on tie end). Note significant alignment deviation at POD.
WOLF
Railway Consulting
61

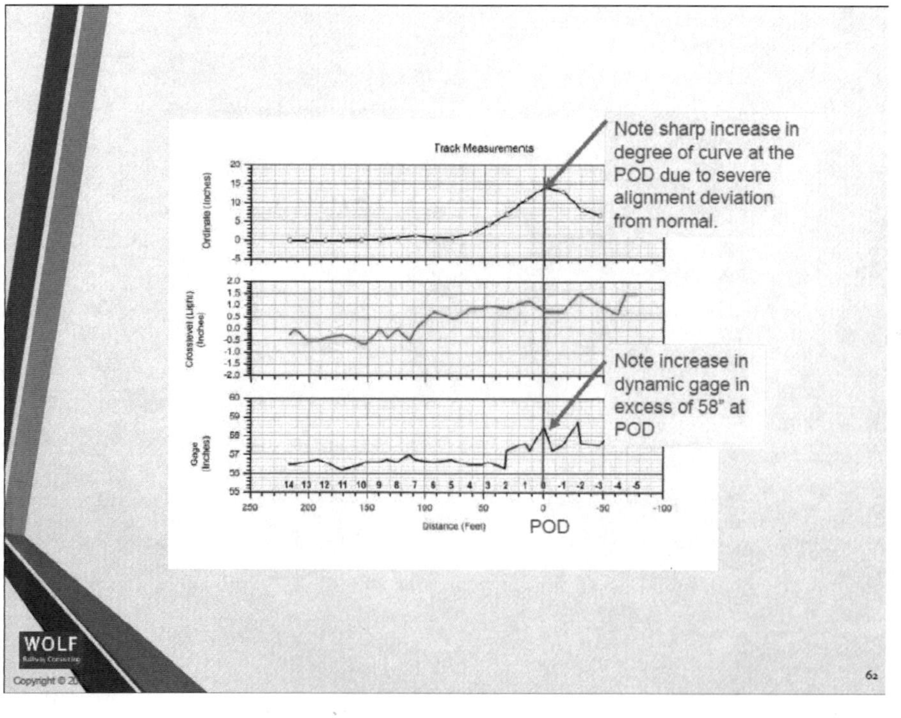

Track Measurements
Ordinate (Inches)
Crosslevel (Light) (Inches)
Gage (Inches)
Distance (Feet)
POD
Note sharp increase in degree of curve at the POD due to severe alignment deviation from normal.
Note increase in dynamic gage in excess of 58" at POD
WOLF
Railway Consulting
62

Truckside L/V Results: Measured vs. Smoothed Alignment

Note, peak L/V Ratios at the POD

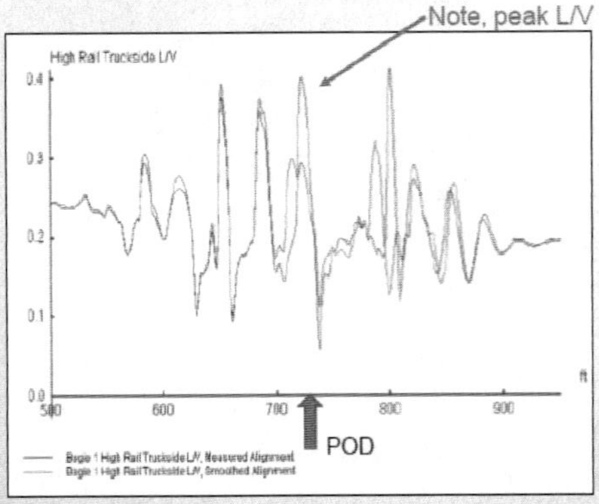

Truck Side L/V at POD:

Measured Alignment = 0.40

Smoothed Alignment = 0.29

WOLF Railway Consulting

63

Poor wheel rail contact geometry

WOLF Railway Consulting

64

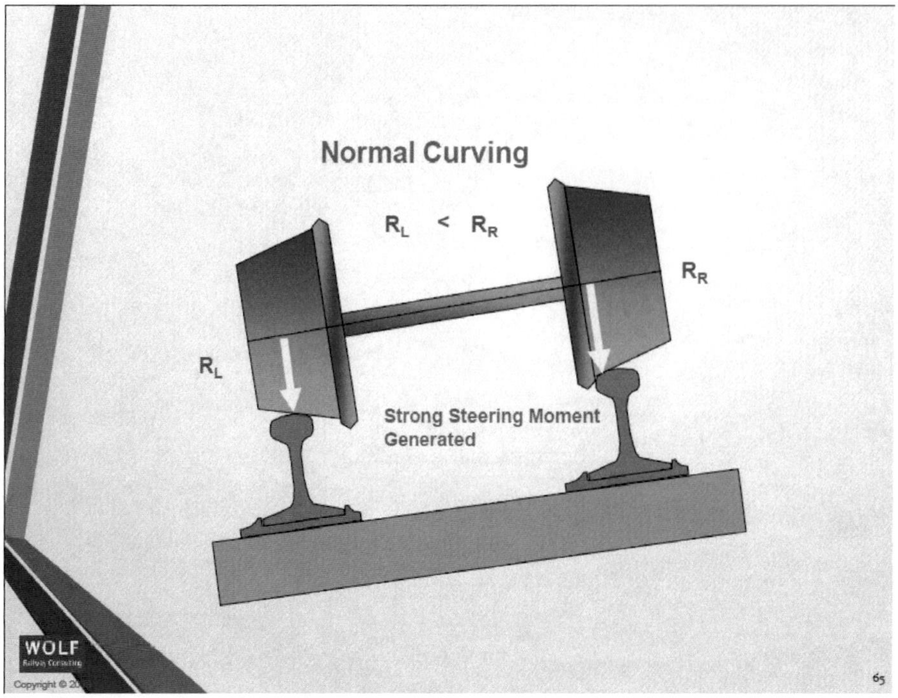

Normal Curving
RL < RR
RR
RL
Strong Steering Moment Generated
WOLF
65

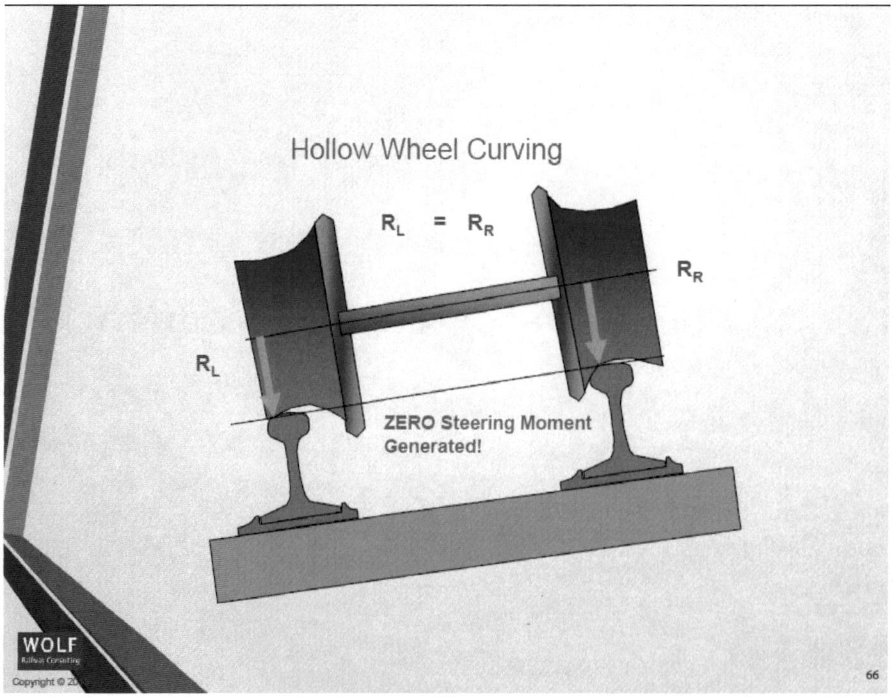

Hollow Wheel Curving
RL = RR
RR
RL
ZERO Steering Moment Generated!
WOLF
66

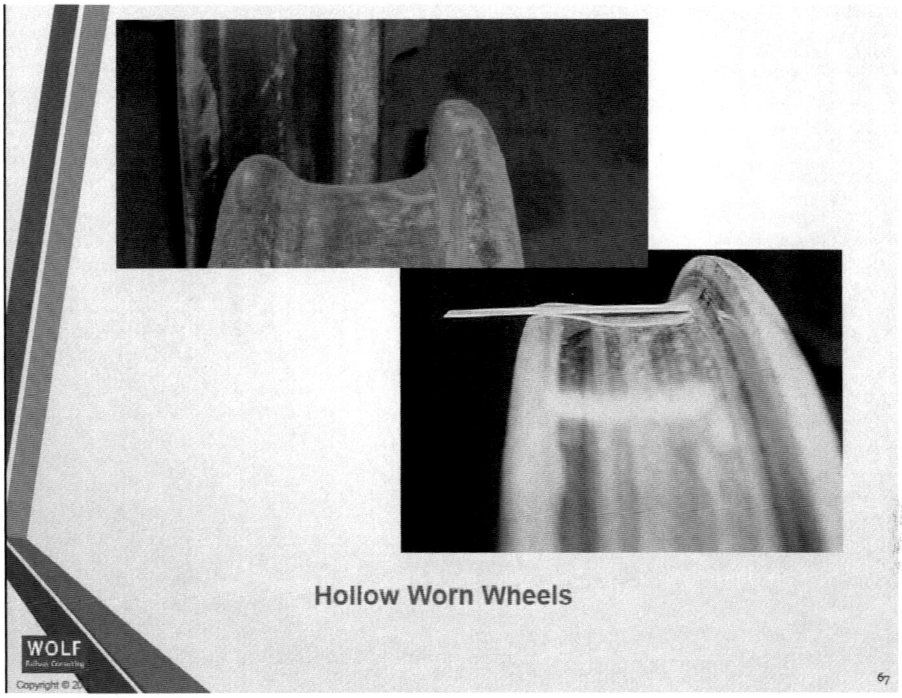
Hollow Worn Wheels
WOLF
67

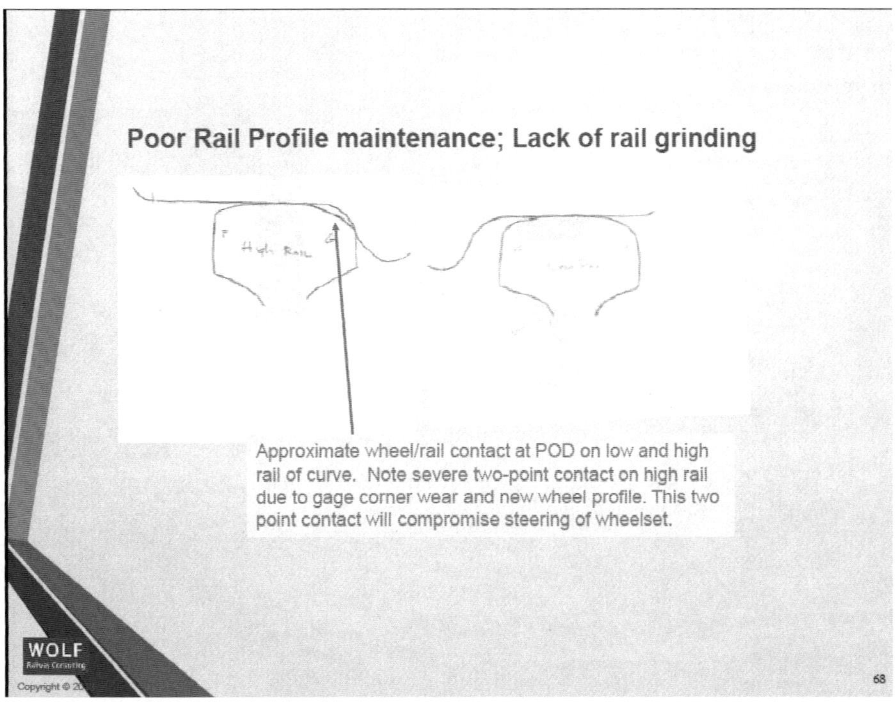
Poor Rail Profile maintenance; Lack of rail grinding
High Rail
Approximate wheel/rail contact at POD on low and high rail of curve. Note severe two-point contact on high rail due to gage corner wear and new wheel profile. This two point contact will compromise steering of wheelset.
WOLF
68

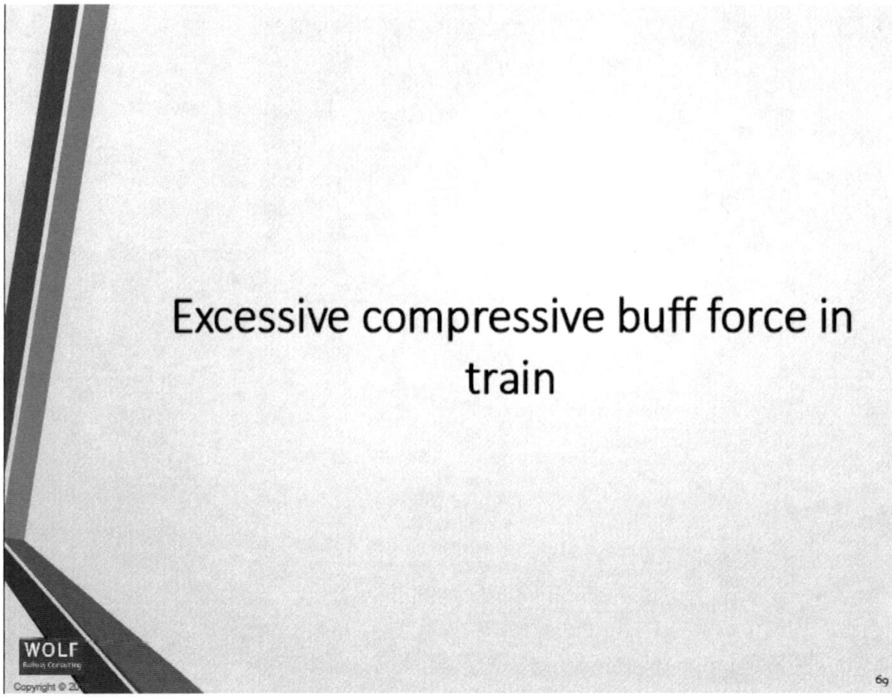
Excessive compressive buff force in train
WOLF
Railway Consulting
69

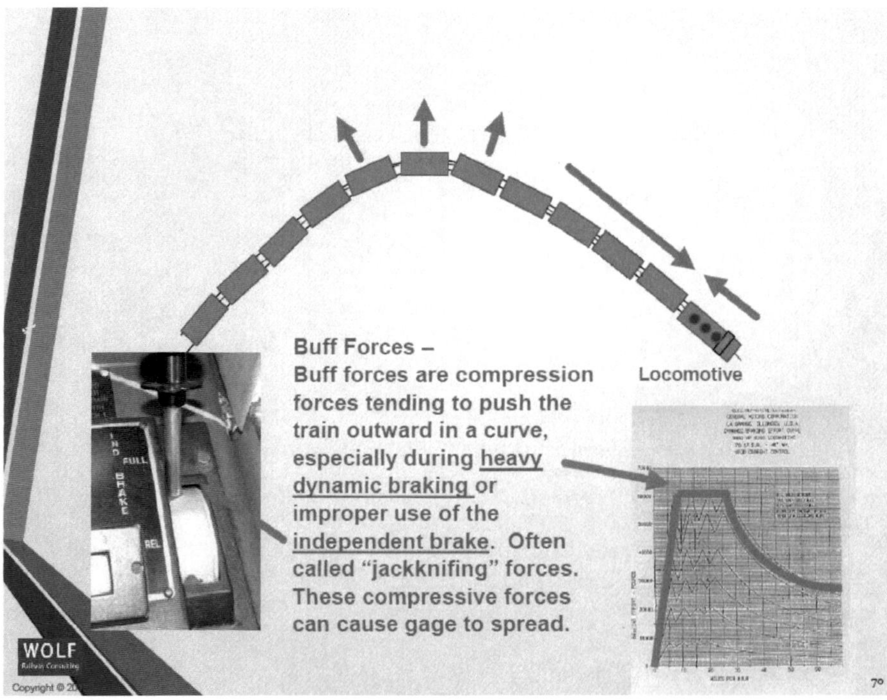
Buff Forces –
Buff forces are compression forces tending to push the train outward in a curve, especially during heavy dynamic braking or improper use of the independent brake. Often called "jackknifing" forces. These compressive forces can cause gage to spread.
Locomotive
FULL
BRAKE
REL
WOLF
Railway Consulting
70

Stiff truck condition; bolster not steering

Truck Bolster Position During Curve Negotiation

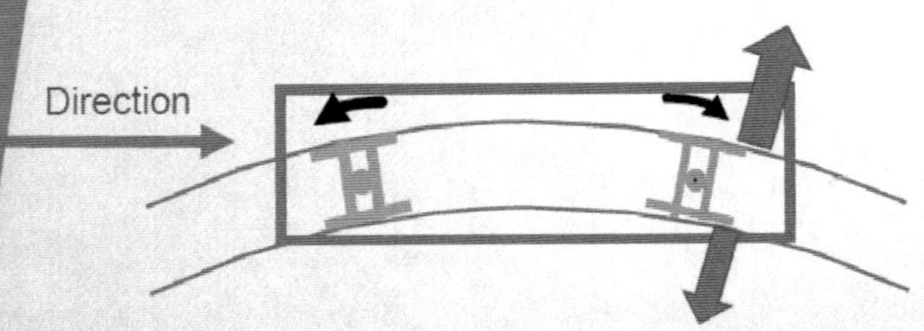

Lead and Trailing Truck Bolsters must pivot in opposite directions.

Lateral forces highest on <u>lead truck</u> during curving

Improper centerplate repair
Lack of bowl lubrication
Insufficient side bearing clearance
Fouling between bowl and rim

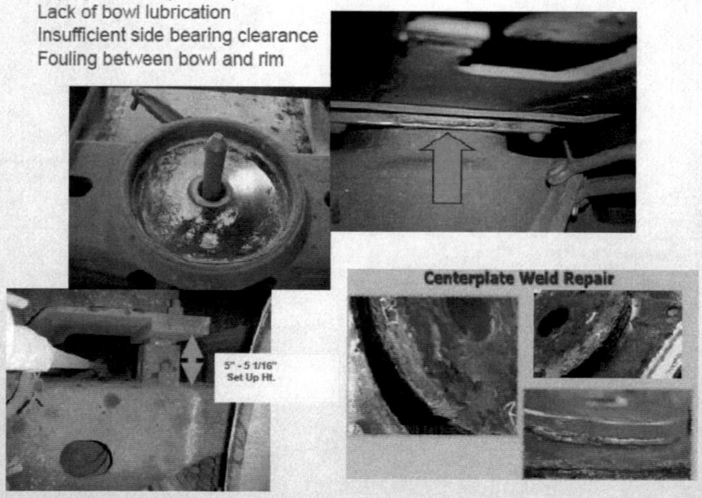

Truck Warp Condition

Truck Warp Restraint

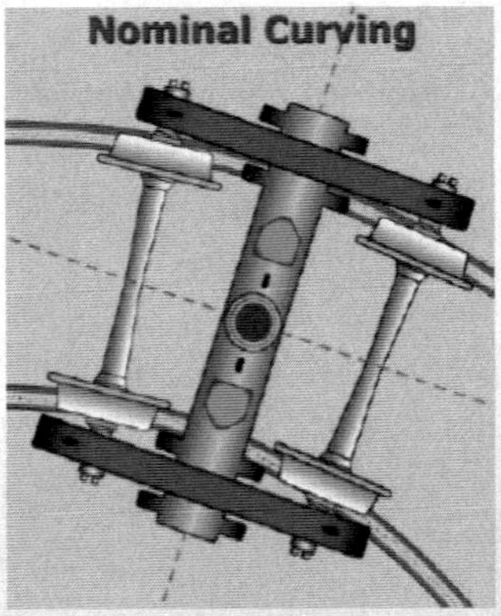

Ideally, a truck should remain "Square" during curving to allow radial alignment of wheelsets with curve

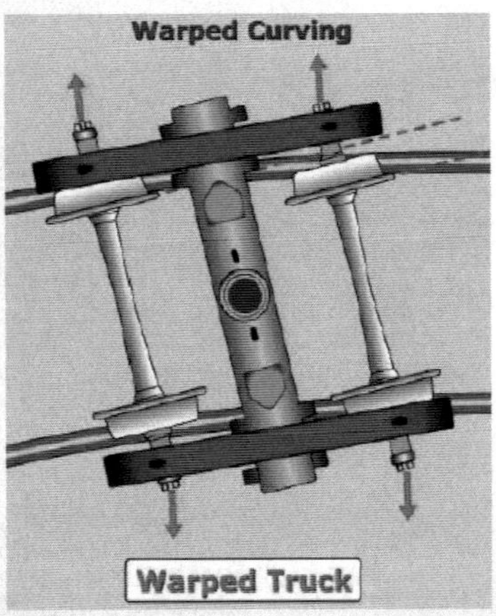

During truck warp, extremely high gage spreading forces are pushing both the high and low rail outward due to high angles of attack on both lead and trail wheelset.

76

OK, we now understand that progressive and catastrophic gage widening is caused by excessive lateral force against both the high and low rail of curves.

Now let's review various track conditions that precipitate widening of the gage....

WOLF

77

Lateral Rail shift to field side

Rail rollover to field side

WOLF

78

Wide track gage is caused by:

- Lateral translation of one or both rails, due to:
  - Deformation of wood fibers holding spikes; elongated spike holes; spike kill
  - Broken screw spikes or cut spike fasteners
  - Worn shoulders on tie plates
  - Gage face wear on the rail
- Rail rolling outwards, due to:
  - Differential tie plate cutting; rail cant
  - Loose fasteners; weak spiking; insufficient spiking
  - Poor wheel contact geometry toward field side of rail head
  - Hollow worn wheels contacting field side of rail head
- A combination of lateral translation and rail roll

WOLF

79

Video courtesy Norfolk Southern

WOLF

80

Video courtesy Norfolk Southern

81

Deformation of track components

82

Spike Killing
Elongated Field side
Spike hole and showing
Loose spike in hole
WOLF
83

Spike Kill
WOLF
84

Raised Spikes
WOLF
Copyright © 20
85

Epoxy Plugging compound
WOLF
Copyright © 20
86

Watch for Tie Seat Abrasion under Rail

Tie Seat Abrasion under Rail and Pad

88

Dynamic Gage Widening

No more than 1/2" lateral movement per FRA

WOLF
Railway Consulting

89

Broken fasteners

WOLF
Railway Consulting

90

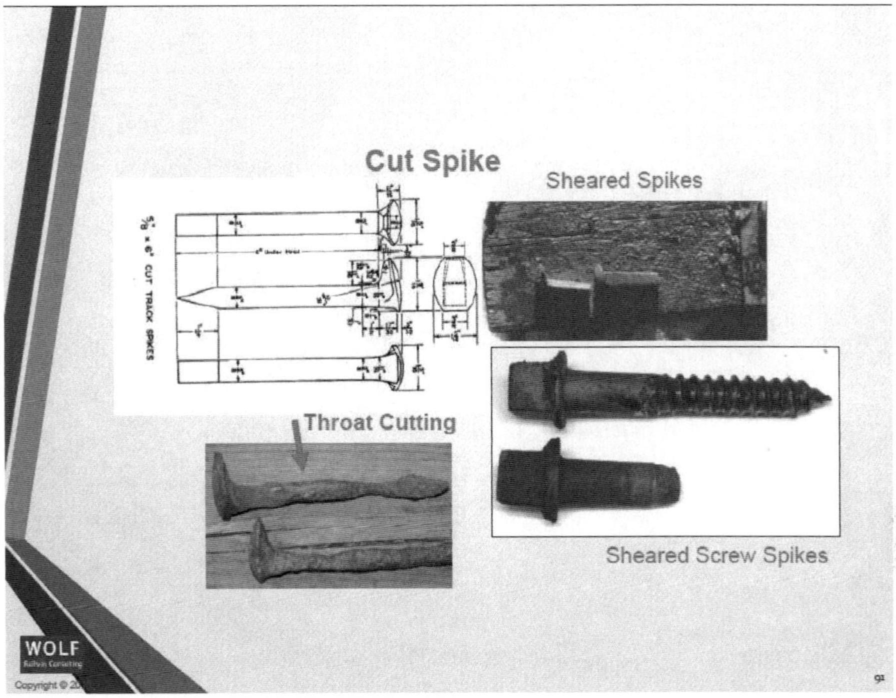
Cut Spike
Sheared Spikes
Throat Cutting
Sheared Screw Spikes
WOLF

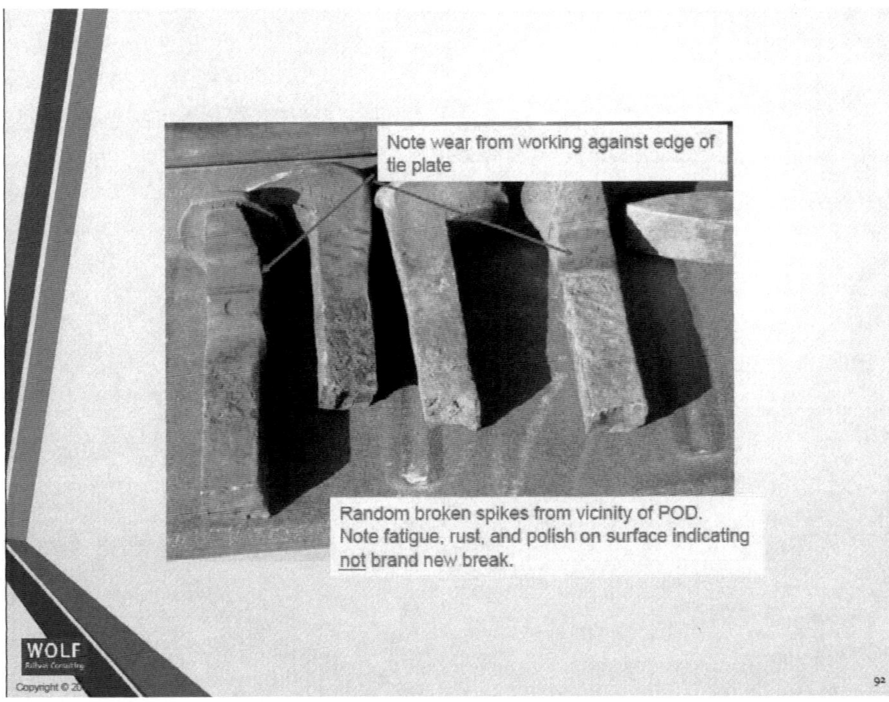
Note wear from working against edge of tie plate
Random broken spikes from vicinity of POD.
Note fatigue, rust, and polish on surface indicating not brand new break.
WOLF

Fracture surfaces on three random broken spikes in Vicinity of POD. Note breaks appear to be "old".

WOLF
Railway Consulting

93

Wide Gage Derailments T110, T111 - Symptoms

- Wheels found dropping in the gage
- May find wheel Drop-in Mark on gage corner of rail at POD
- Pre-existing Wide Gage or Evidence of Dynamic Gage Widening
- Either Rail May be Rolled Outward

WOLF
Railway Consulting

94

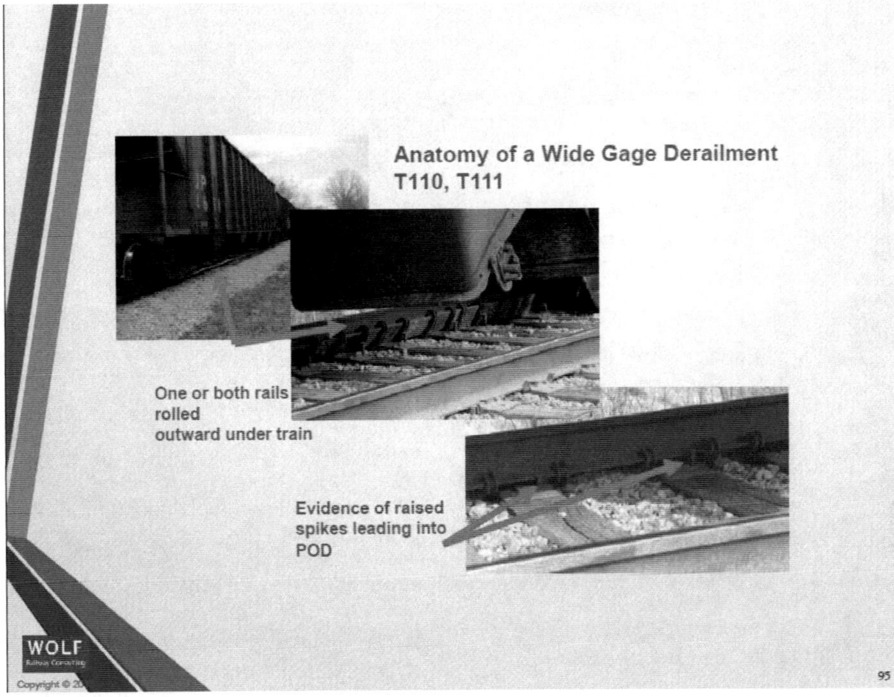
Anatomy of a Wide Gage Derailment
T110, T111
One or both rails rolled
outward under train
Evidence of raised spikes leading into POD
WOLF
95

Wheel Drop In Mark due to rail rolling outward
WOLF
96

Wide Gage Summary

- Wide gage derailments involve complex interaction between the track, rail, vehicles, and operating parameters.
- The investigator must determine if the gage widened progressively or catastrophically.
- Each component of the system must be evaluated
  - Ties, fasteners, wheel-rail contact, wheelset steering, rail wear, track geometry, speed, wheel wear, vehicle condition, operating factors
- Only after a detailed investigation can a true root cause(s) be developed and corrective actions be implemented.
- Remember: Not all wide gage derailments are due to weak track!

WOLF Railway Consulting

97

Turnout Derailments

Split Switches....

Thin flange wheels...

Broken Points....

Run through Switches...

WOLF Railway Consulting

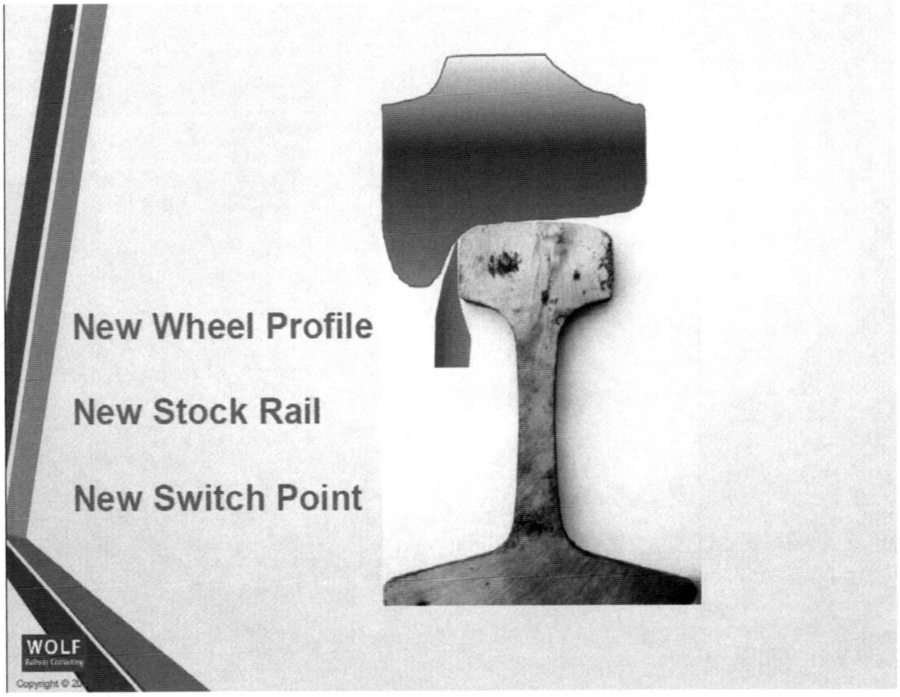
New Wheel Profile
New Stock Rail
New Switch Point
WOLF

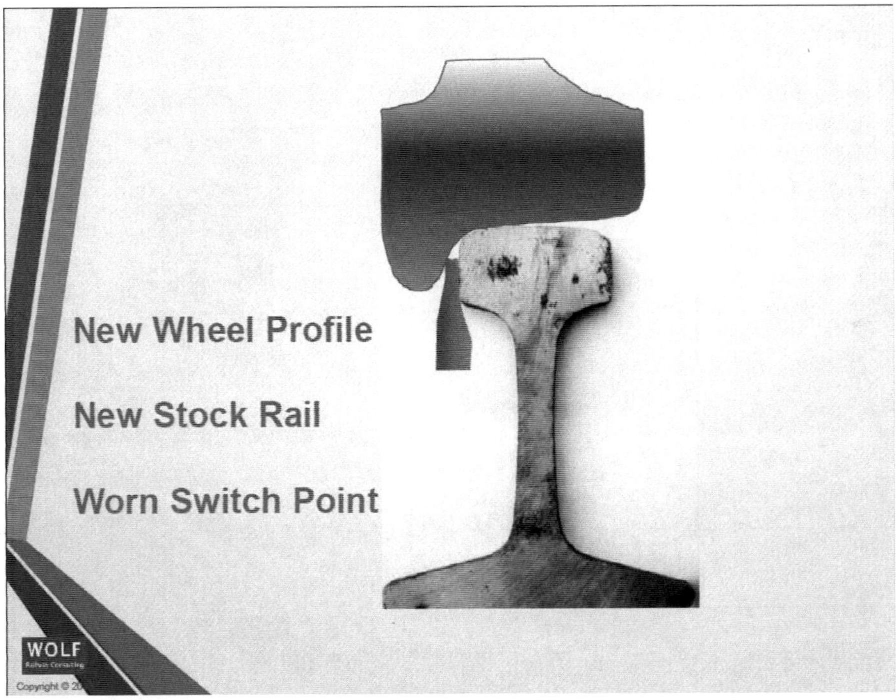
New Wheel Profile
New Stock Rail
Worn Switch Point
WOLF

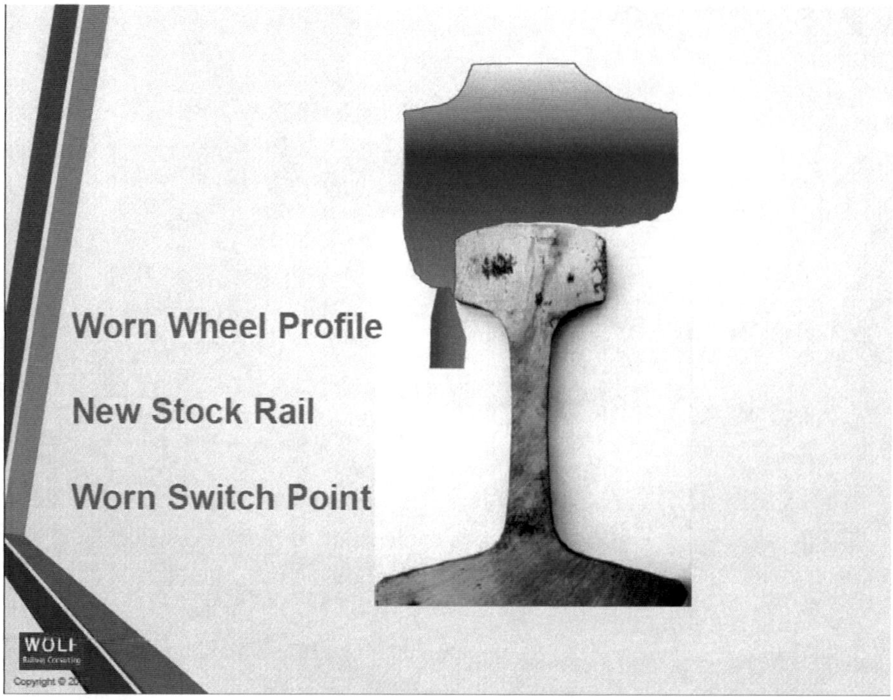
Worn Wheel Profile
New Stock Rail
Worn Switch Point
WOLF
Railway Consulting

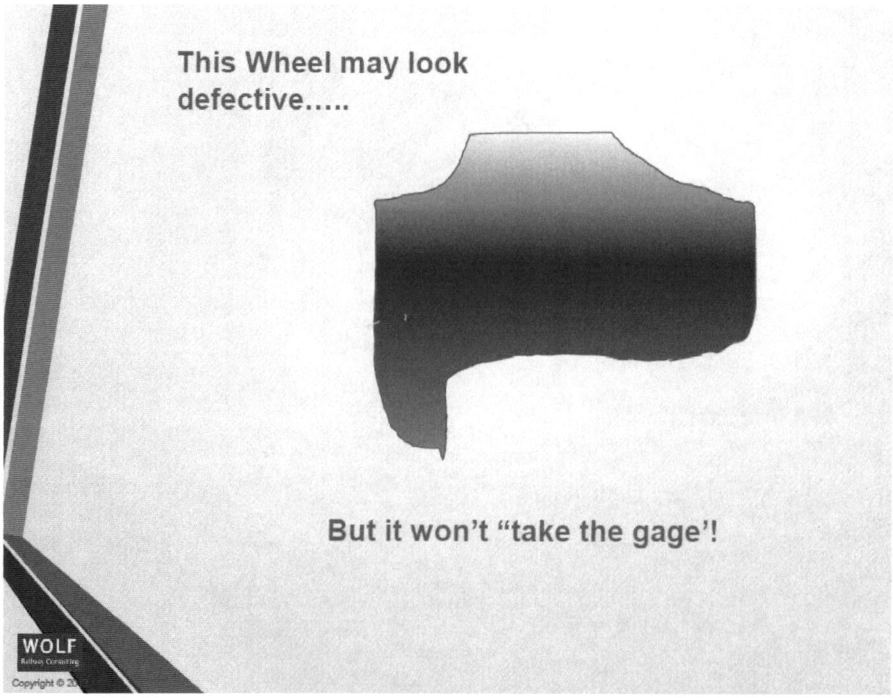
This Wheel may look defective.....
But it won't "take the gage'!
WOLF
Railway Consulting

Asymmetrical Wheel Wear (left and right) is quite common

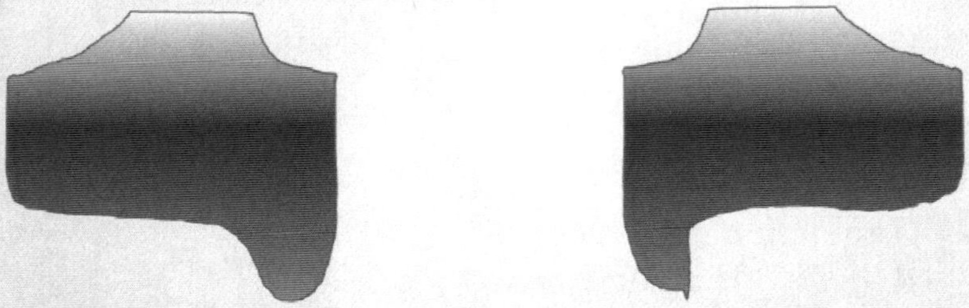

This can be an indication that truck is not steering properly in one direction due to stiff truck or truck warp condition

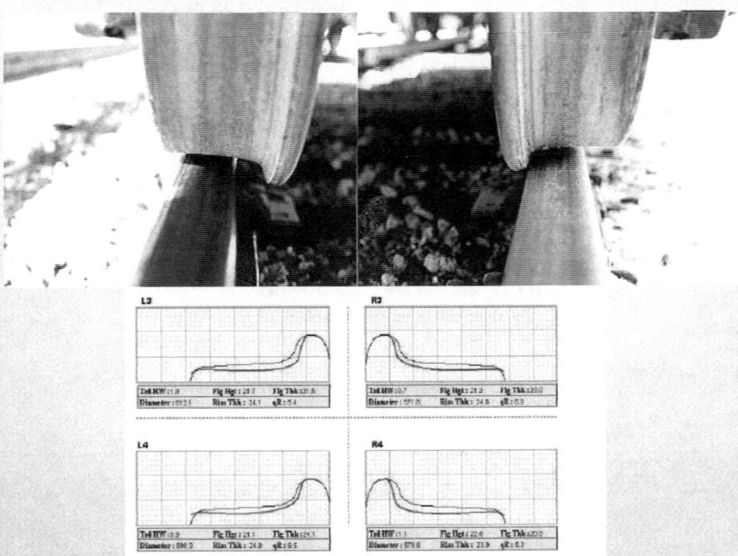

Asymmetrical Wheel Wear

Asymmetrical Wheel Wear can be due to:

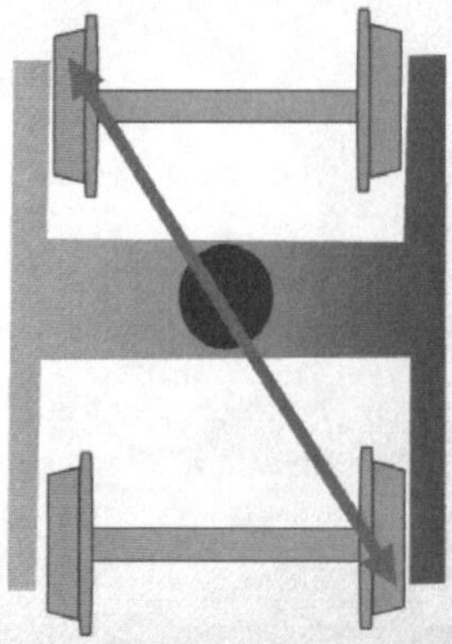

- -Stiff Truck in one direction
- -Tight Side Bearing one side
- -Shifted Load
- -Centerbowl Contact
- -Mismatched side frames
- -Mismatched Wheel tapes
- -Fouling of Centerbowl rim
- -Uneven Braking forces

Bottom line... Uneven Wheel Wear is often a symptom of truck turning problem. Investigator Must find reasons for truck turning problem.

Note: sometimes Uneven wheel wear is a symptom of uneven braking forces due to rigging or brake beam angularity.

The Question is...

- Is the switch point the cause due to wear and chipping?
  - CFR 213.135 (h) – "Unusually chipped or worn switch points shall be repaired or replaced..."
- Or is the wheel/truck the cause due to thin flange wheel or stiff truck condition?
  - AAR Rule 41 A. – "Thin flange 15/16" or less condemnable at any time"

Investigating Switch Point Derailments

- Inspect point for wear, fit, gapping
- Inspect point for vertical movement due to loose heel block
- Inspect truck as a system for tight side bearings, wedge rise, bowl contact, button mismatch, etc.
- Check geometry approaching and leaving point area for Crosslevel twist
- Check stock rail braces for gapping or looseness
- Check head block ties for rot, decay, pumping, gage issues
- Check switch point throw
- Inspect switch rods for deterioration, loose bolts, cotters, etc.
- Check switch machine for excess play or loose spiking

WOLF

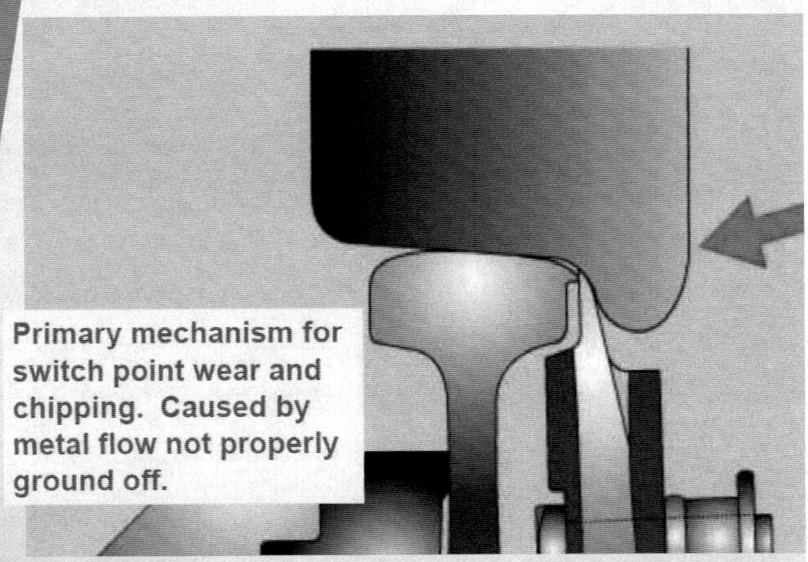

WOLF

Switch Point Issues
WOLF
Copyright © 20

Switch Point Issues
Copyright © 20

WOLF
Gapped Switch Point

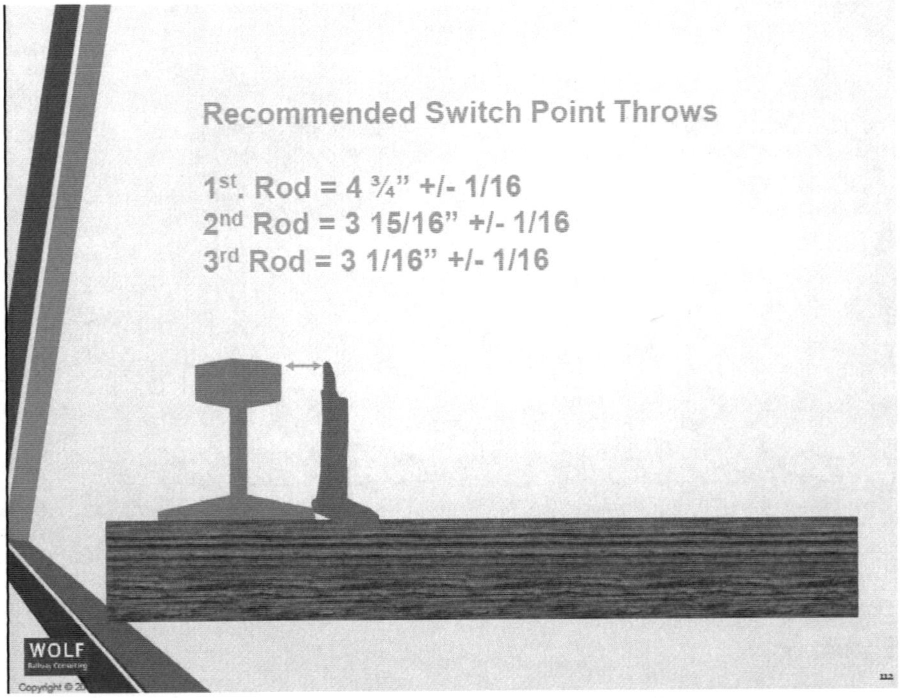
Recommended Switch Point Throws
1st. Rod = 4 ¾" +/- 1/16
2nd Rod = 3 15/16" +/- 1/16
3rd Rod = 3 1/16" +/- 1/16
WOLF

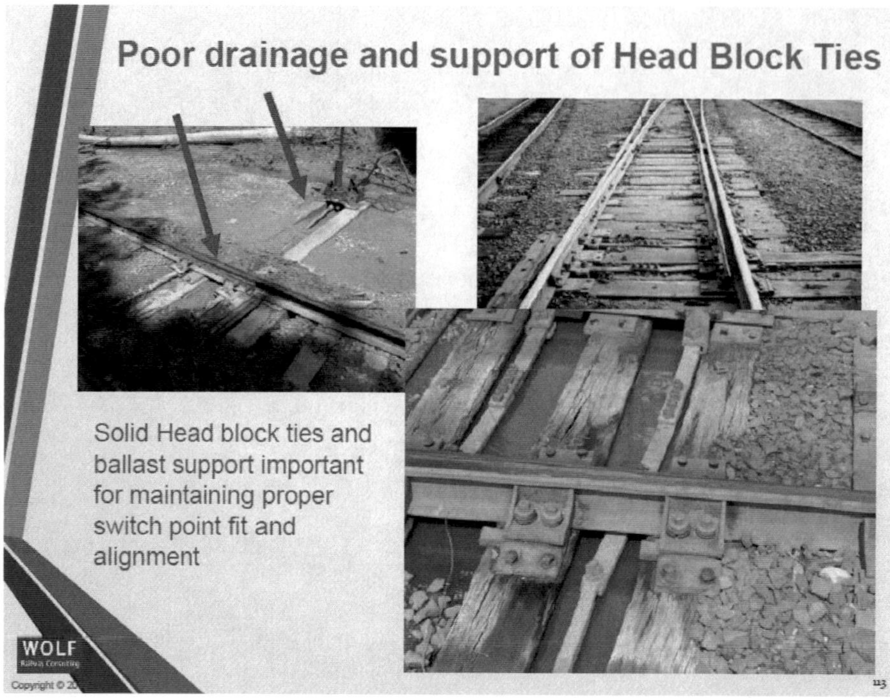
Poor drainage and support of Head Block Ties
Solid Head block ties and ballast support important for maintaining proper switch point fit and alignment
WOLF
113

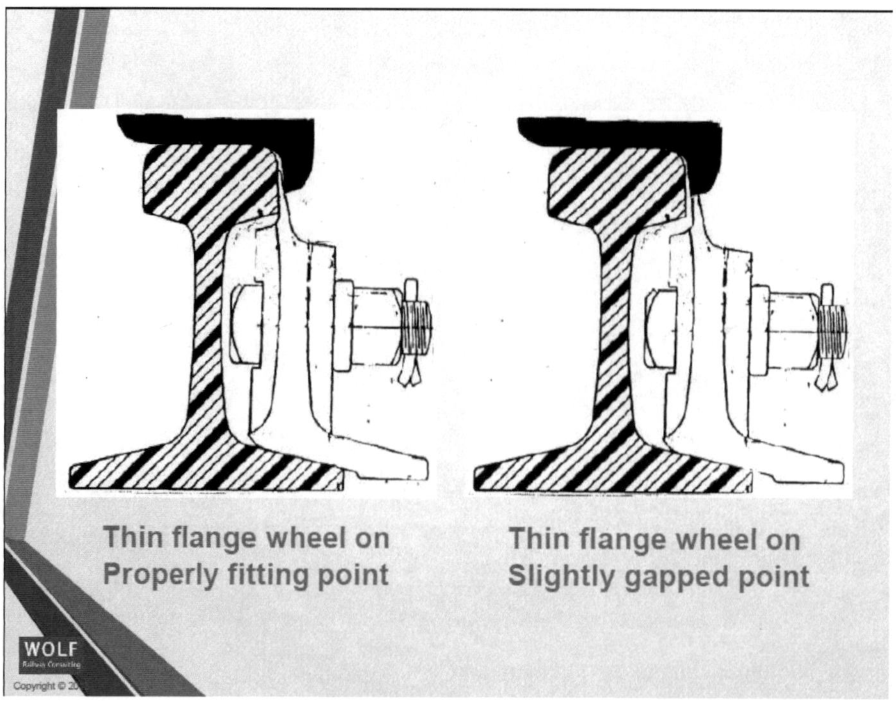
Thin flange wheel on Properly fitting point
Thin flange wheel on Slightly gapped point
WOLF

Heel Block;
Watch for Vertical Pumping and
Gapping at Point under Motion

Heel Block Pumping under wheel load can lead to point raising vertically

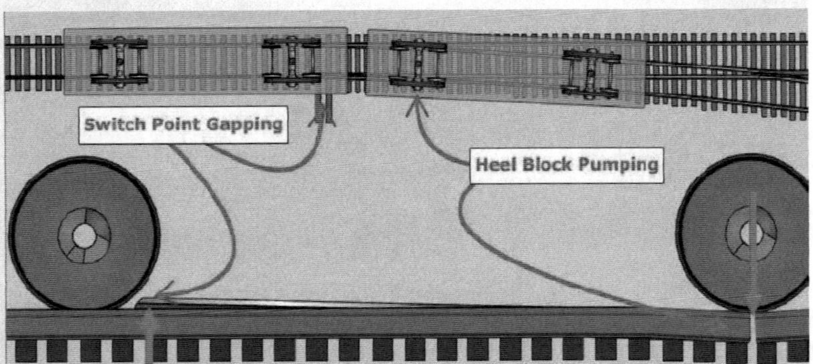

Loose & Missing
Heel Block Bolts
WOLF

Loose & Missing
Heel Block Bolts

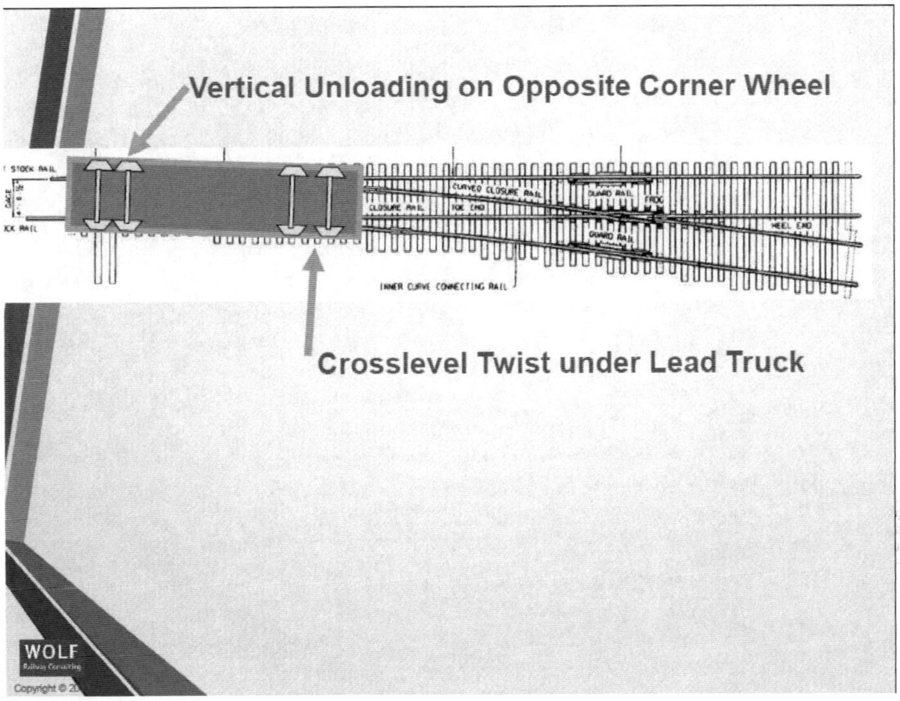
Vertical Unloading on Opposite Corner Wheel
Crosslevel Twist under Lead Truck
WOLF

Potential for twist as lead looses elevation
WOLF

Potential for twist as lead looses elevation
WOLF

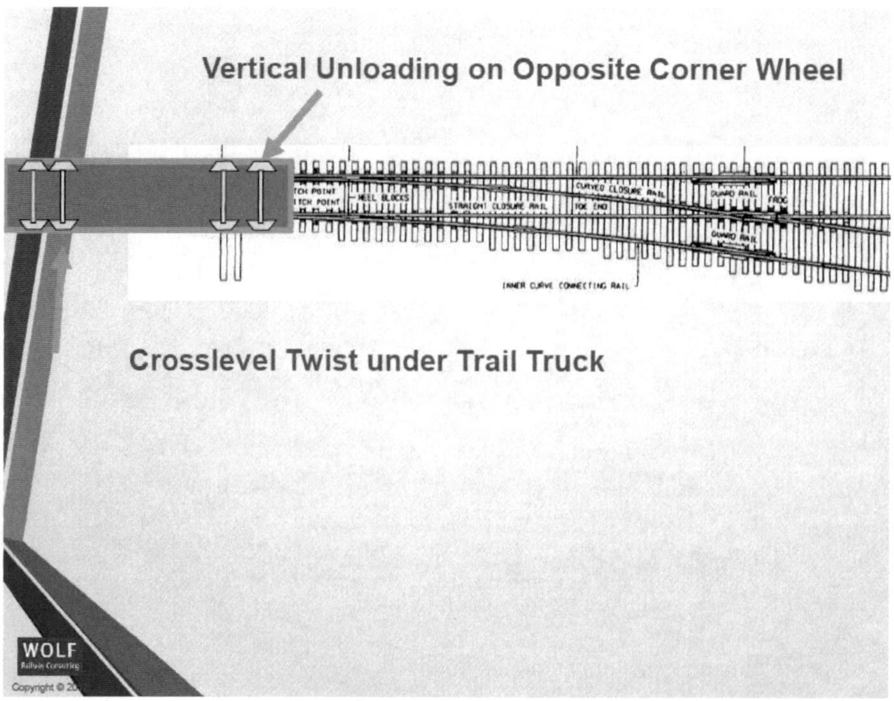
Vertical Unloading on Opposite Corner Wheel
Crosslevel Twist under Trail Truck
WOLF

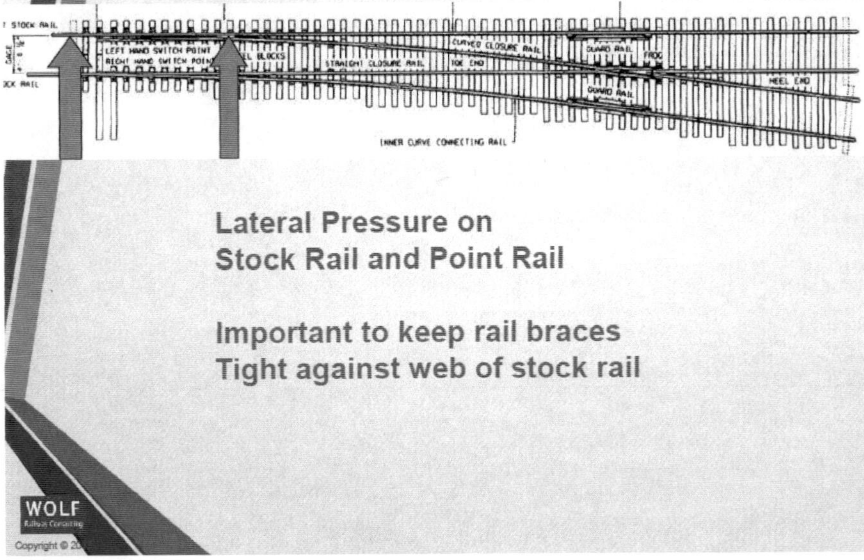

Lateral Pressure on
Stock Rail and Point Rail

Important to keep rail braces
Tight against web of stock rail

Rail Braces; Keep tight
WOLF
Copyright © 20

Loose Rail
Braces
WOLF
Copyright © 20

Summary - Switch Derailments

- Must look beyond wheel flange; inspect entire car & trucks as a system
- Must look beyond switch point; inspect entire switch as a system
- Also include human factor issues
  - High Buff/Slack Forces
  - Run Through Switch
  - Improper Handling & Inspection of Switch

WOLF

Track Geometry Issues in Yards

- Crosslevel Twist or Variance
  - Check twist in the 40' to 62' range (truck center distance)
  - Compliance with FRA/TC standards does not always insure there will be no derailment
    - 3" X-level and 79% defective ties OK???
    - Combinations of compliant twist, alignment, profile at same point in track
    - Yard Drainage often cause of crosslevel twist
- Wide Gage and Alignment covered already
  - Back to Back turnouts and reversing curves
  - Need at least one car length tangent if possible
- Rail Joints
  - Proper support and TIGHT bolts
- Flangeway Clearance
  - Crossings and turnouts
  - Remove dirt, mud, gravel, ice, etc.

WOLF

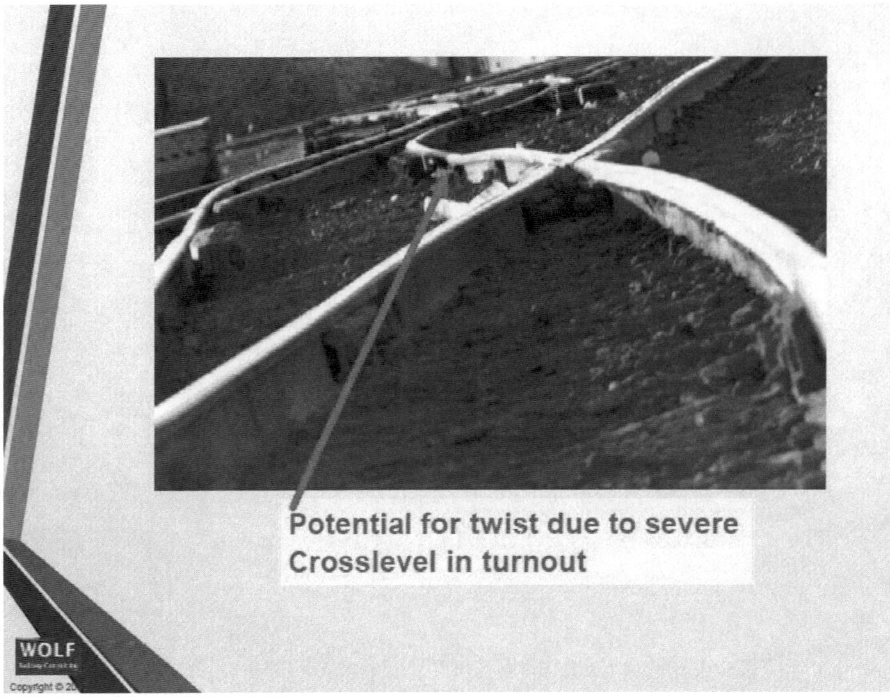
Potential for twist due to severe
Crosslevel in turnout
WOLF

Insufficient Intervening tangent between turnouts
WOLF
130

Severe alignment kink in turnout
WOLF
131

Poor joint support and bolting
WOLF
132

Heavily fouled flangeway in crossing

WOLF

133

Mechanical Causes

- **Side bearing clearance or set up height**
  - Standard Roller or Block side bearings
  - Constant Contact side bearings
  - Cars with tight sidebearings cannot accommodate FRA/TC compliant track twist conditions
- **Friction Casting Wedge Rise**
  - AAR Rule 46
  - Friction casting wedge rise can be cause of truck warp
- **Stiff Truck Condition**
  - Fouling of centerbowl rim
  - Improper centerplate repair

WOLF

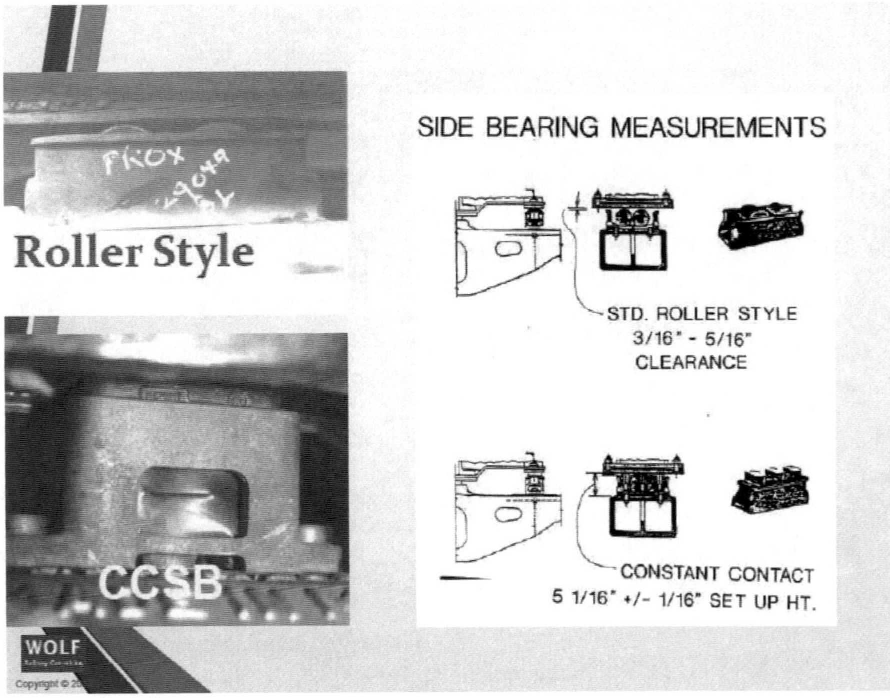
Roller Style
SIDE BEARING MEASUREMENTS
STD. ROLLER STYLE
3/16" - 5/16"
CLEARANCE
CCSB
CONSTANT CONTACT
5 1/16" +/- 1/16" SET UP HT.
WOLF

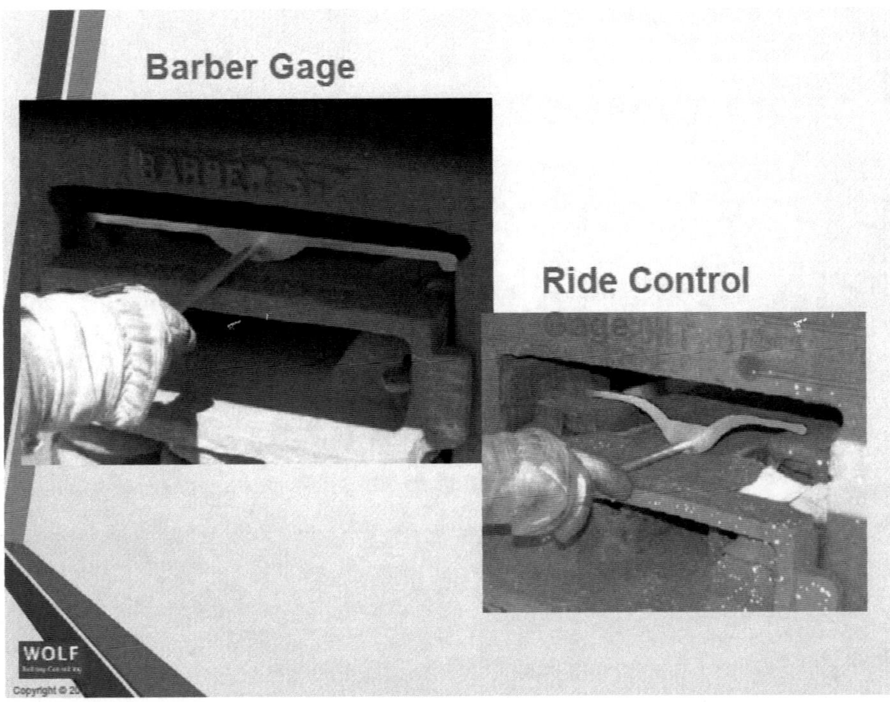
Barber Gage
Ride Control
WOLF

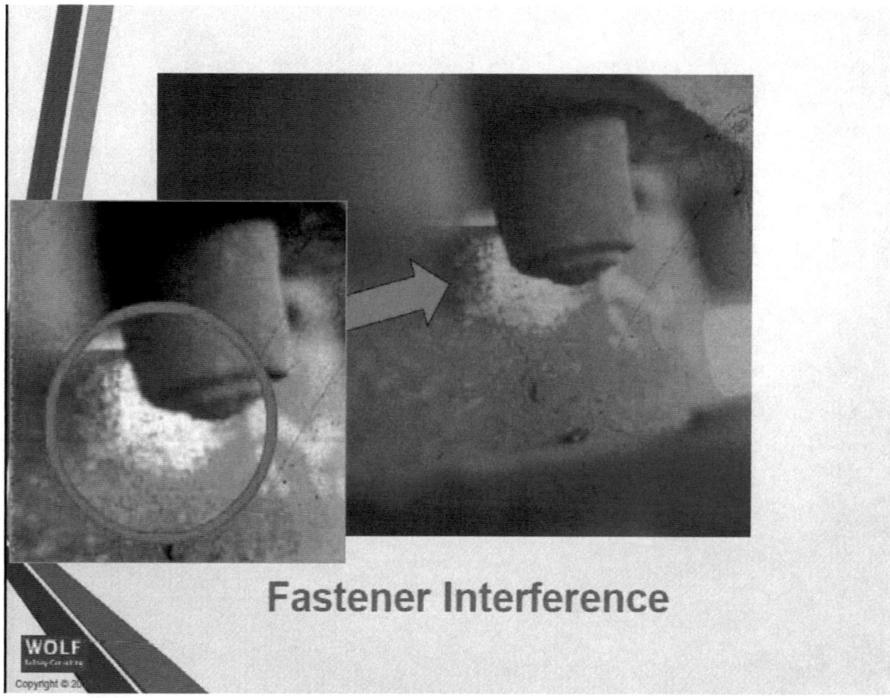
Fastener Interference
WOLF

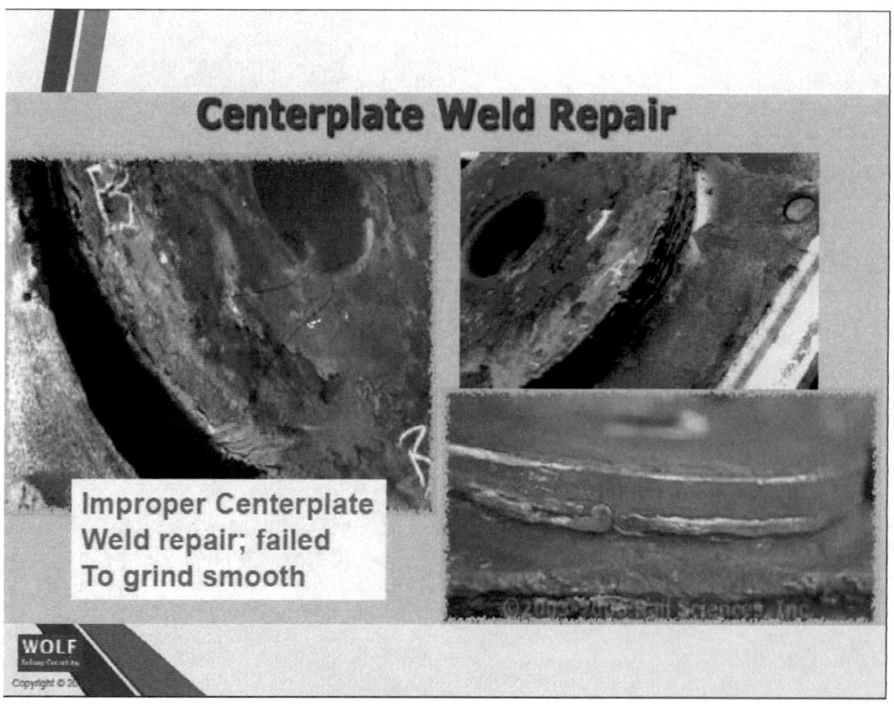
Centerplate Weld Repair
Improper Centerplate
Weld repair; failed
To grind smooth
WOLF

Conclusions – Yard & Terminal Derailments

- Constant Focus on Human failures; reportable and non-reportable
- Rules testing and compliance mandatory
- Wide gage track is not always caused by a track deficiency; check mechanical and human factors
- Thorough inspection of equipment is mandatory; only way to find certain mechanical defects is to jack car and inspect centerplate area

WOLF

The End

WOLF

Joel Morley, Transportation Safety Board of Canada

Joel Morley is a Senior Human Performance analyst at the Transportation Safety Board of Canada, a role he filled from 2001 to 2007 and returned to in 2012. While at the TSB, Joel has participated in numerous investigations in the air, rail and marine modes and trained a large number of investigators in the theory and practice of investigating for human performance factors. In addition to the TSB, Joel has experience applying safety management and human factors for a large air navigation service provider and has worked as a researcher and consultant. Joel holds a PHD in Applied Psychology from Cranfield University.

Investigating for Organizational and Management Factors

Joel Morley
IAROO – 22 September 2014

Canadä

Organization & Management Factors

Conditions which:

- Negatively impact human performance and make errors more likely; or,
- Negatively impact the organization's ability to proactively identify and deal with these conditions.

Finding O&M Factors

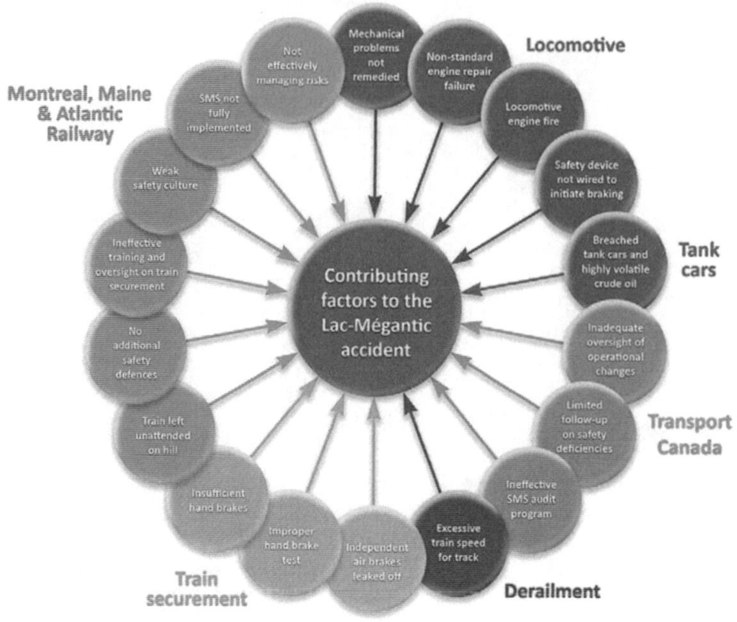

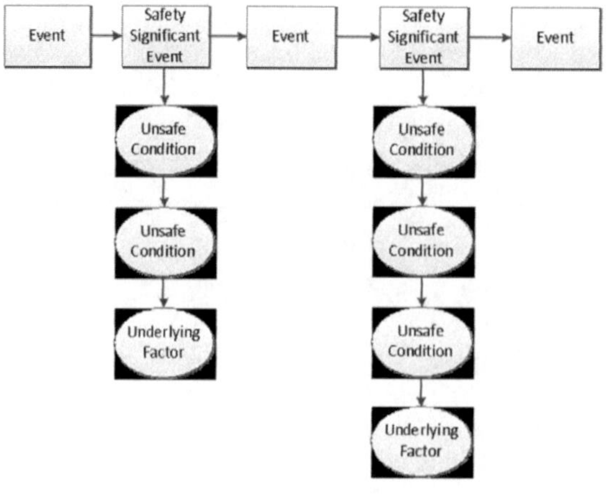

Transportation Safety Board of Canada Bureau de la sécurité des transports du Canada

6

Background

- O&M Investigation Guide first produced in 2001
- Module in Human Factors Course
- Increasing regulatory requirement for SMS
- Number of investigations with significant safety management focus
- Considerable learning within the organization
- Areas for improvement remain – led to development of new guide and tools

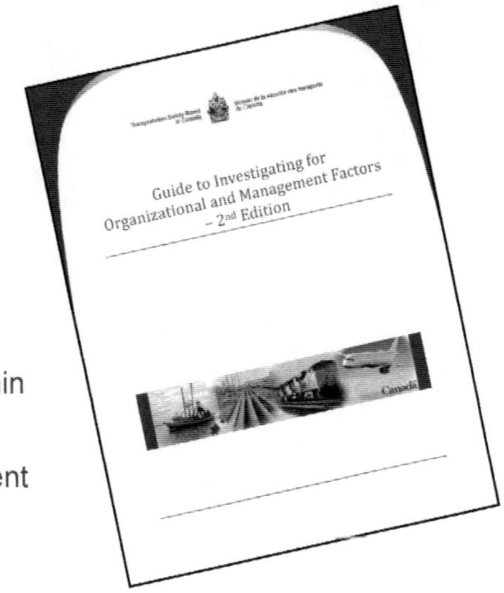

Transportation Safety Board of Canada   Bureau de la sécurité des transports du Canada

7

Method

- Met with Board, sample of Managers, IICs
- Reviewed selection of reports for Board comments
- Developed list of challenges
- Met with NTSB colleagues
- Review of literature
- Developed Guide and Toolkit (targeted to challenges)
- Piloted tools internally

Transportation Safety Board of Canada   Bureau de la sécurité des transports du Canada

8

Challenges

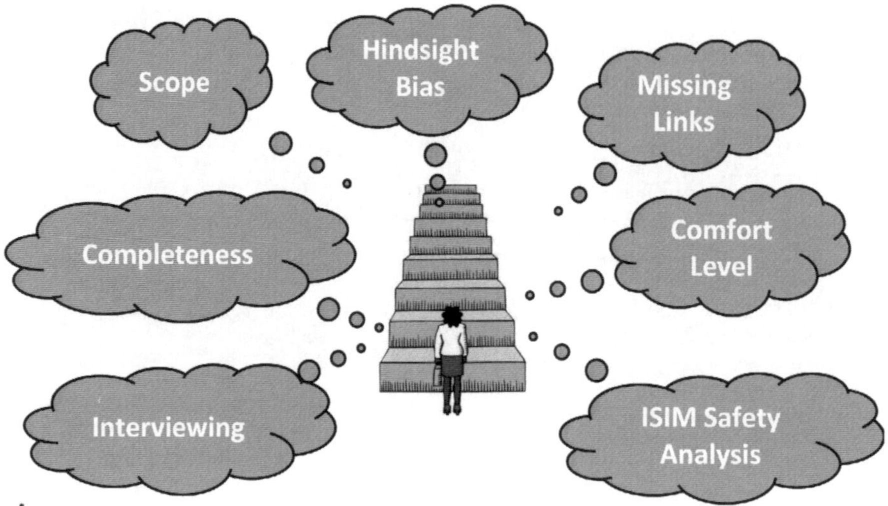

Transportation Safety Board of Canada   Bureau de la sécurité des transports du Canada

9

Constructs and Definitions

- Safety
- Hazard, Risk and ALARP
- Latent unsafe conditions
- safety management
- Safety Management System
- Safety Culture

Transportation Safety Board of Canada   Bureau de la sécurité des transports du Canada

10

Safety

Safety is the state in which the risk of harm to persons or property damage is reduced to, and maintained at or below, an acceptable level through a continuing process of hazard identification and risk assessment.

ICAO Safety Management Manual, DOC 9859.

Hazard, Risk, and ALARP

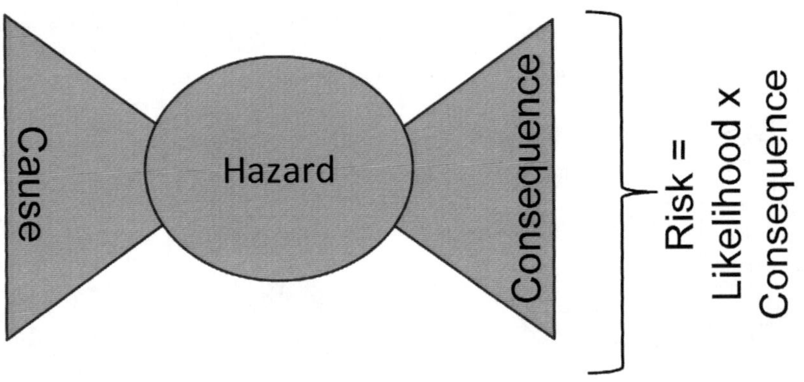

Latent Unsafe Conditions

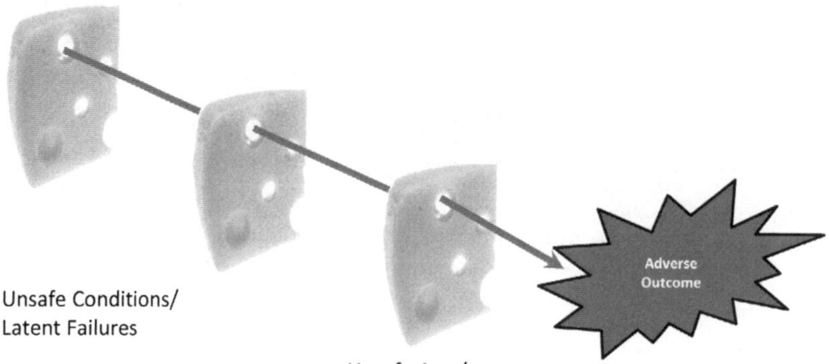

Transportation Safety Board of Canada   Bureau de la sécurité des transports du Canada

13

safety management and SMS

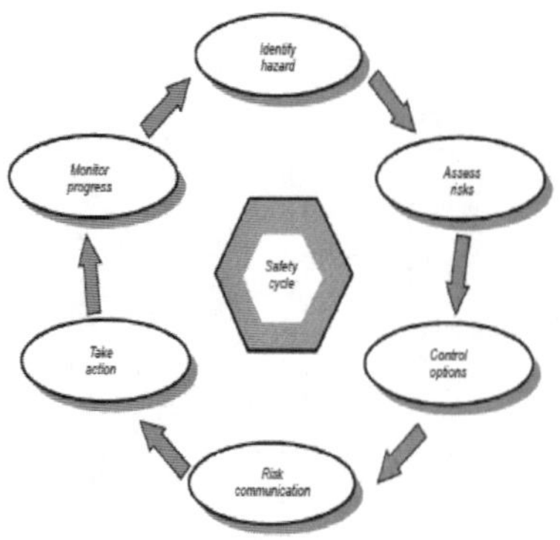

From ICAO Doc 9859.

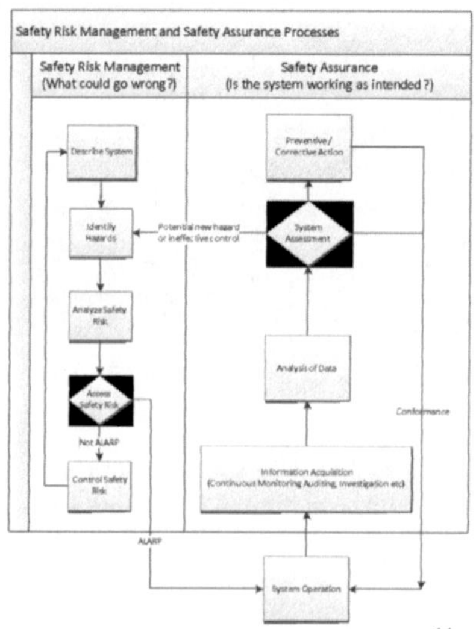

From: Stroltzer et al (2008). Safety Management Systems in Aviation.

Transportation Safety Board of Canada   Bureau de la sécurité des transports du Canada

14

Safety Culture

	Pathological	Burocratic	Generative
Information	Hidden	Ignored	Sought
Messengers	Shouted	Tolerated	Trained
Responsibilities	Shirked	Boxed	Shared
Reports	Discouraged	Allowed	Rewarded
Failures	Covered up	Merciful	Scrutinized
New ideas	Crushed	Problematic	Welcomed
Resulting Organization	**Conflicted**	**"Red-tape"**	**Reliable**

Originally from Westrum (1992). Described in Reason (1997). Managing the Risks of Organizational Accidents.

Transportation Safety Board of Canada / Bureau de la sécurité des transports du Canada

15

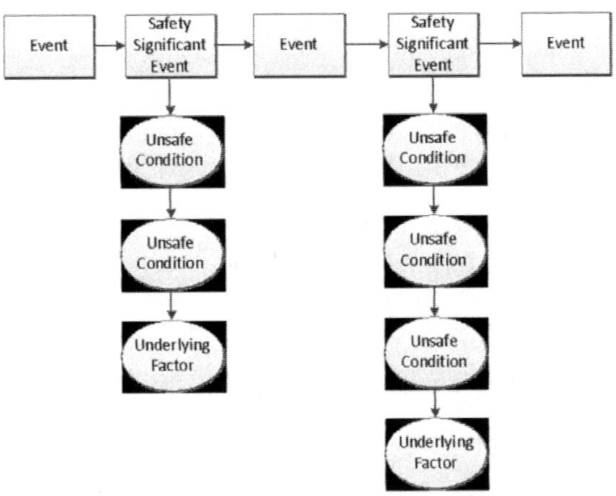

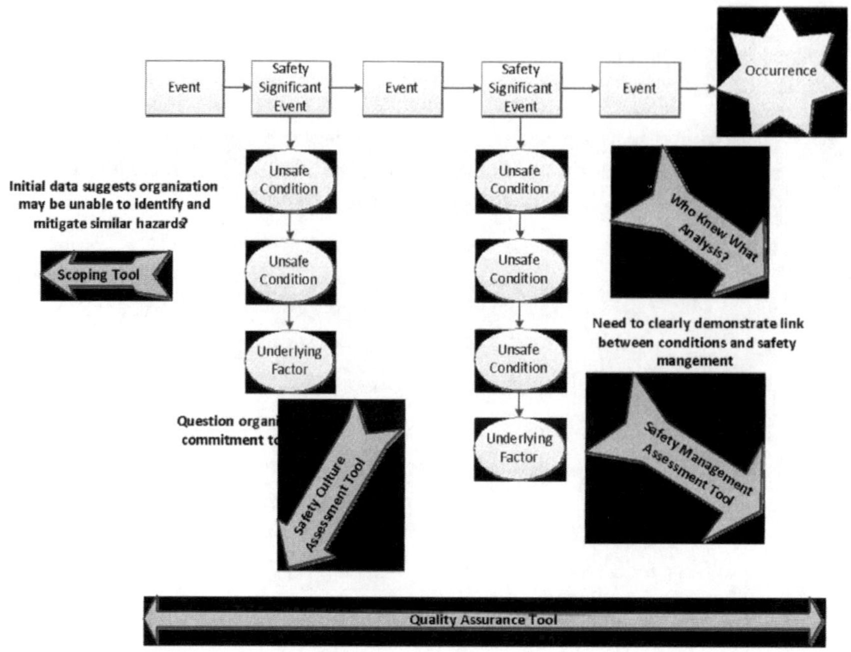

Scoping Tool

- Indicators:
  - Hazards known to some members of the organization before the occurrence
  - Organization ill equipped to identify similar hazards or mitigate similar risks?
  - Disconnect between practices observed to date and documented procedures?
  - Organization known to the regulator or to TSB to be in need of increased oversight?
  - Significant recent transition?
  - Indications that regulatory oversight is lacking?
- Decision → Next Steps → Resources

Transportation Safety Board of Canada Bureau de la sécurité des transports du Canada

18

Scoping Tool

Scoping Tool Worksheet		
Completed by:		Date:
Briefly describe the unsafe acts, unsafe conditions and underlying factors (hazards) identified to date in the occurrence. •		
Scoping Question	*Yes/No*	*Details/Examples*
1. Are there indications that hazards which contributed to the occurrence were known to some members of the organization before the occurrence?		
2. Are there underlying factors identified in the early stages of the investigation which are likely to make it more difficult for the organization to identify similar hazards or mitigate similar risks?		
3. Is there a disconnect between practices observed to date and documented procedures?		
4. Is the organization one which is known to the regulator or to TSB to be in need of increased oversight?		
5. Has the organization or industry gone through significant recent transition?		
6. Are there indications that regulatory oversight is lacking?		
Decision: Describe the scope of the O&M investigation (i.e. preliminary description of issues to be pursued) or reason for limiting scope.		
1.		
Next Steps: Describe the information which will be gathered and by what means.		
1.		
Resources: Describe the resources required to carry out the investigation.		
1.		

Transportation Safety Board of Canada
Bureau de la sécurité des transports du Canada

19

"Who Knew What?" Analysis

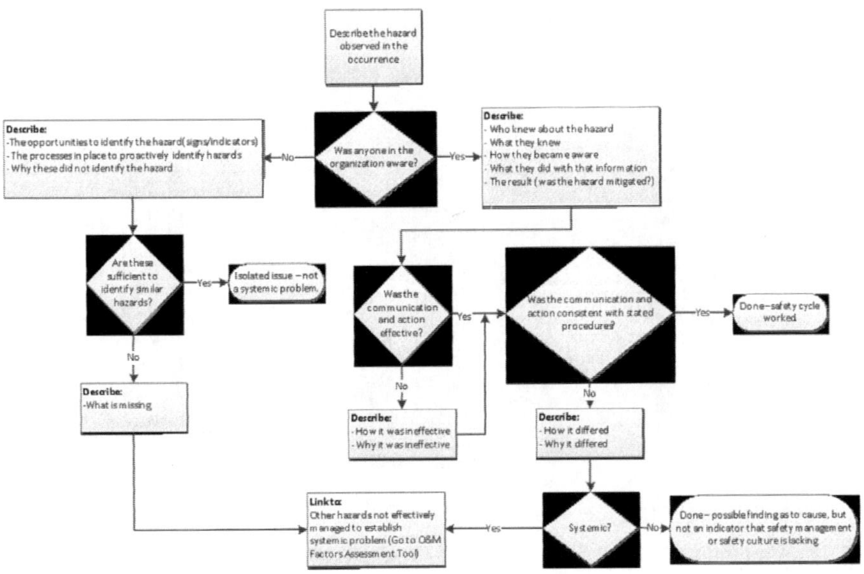

"Who Knew What?" Analysis

Who Knew What? Analysis Worksheet		
Completed by:		**Date:**
Hazard:		
Question	***(Yes/No)***	***Supporting Information***
Was anyone in the organization aware of the hazard?		
Are the processes in place to identify hazards sufficient to identify similar hazards?		
Was communication and action with respect to hazard reasonable and effective?		
Was communication and action with respect to hazard consistent with company safety management procedures?		
Is the handling of this hazard unique and not indicative of a systemic problem?(i.e. the safety management structures are present and generally effective was missing or ineffective?		
Conclusion: What is the picture the information above provides? If there is a systemic issue, describe it briefly (i.e. what safety management structure absent or ineffective; how did actions differ from stated safety management procedures and why?) .		
1.		

Note: Remember to update the ISIM safety analysis with any changes or additions.

O&M Factor Assessment Tool

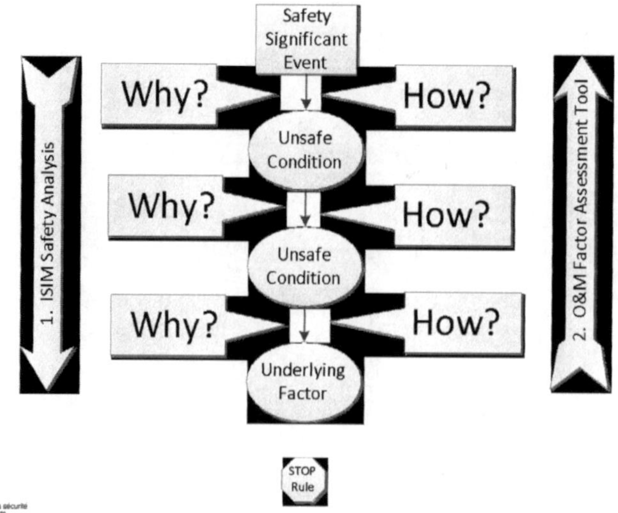

Transportation Safety Board of Canada
Bureau de la sécurité des transports du Canada

22

O&M Factor Assessment Tool

O&M Factor Assessment Tool Worksheet	
Completed by:	**Date:**
O & M Factor (Underlying factor from ISIM Safety Analysis):	
·	
Describe the unsafe acts or conditions to which the actor contributed.	**Describe the link to this occurrence** (***How*** did this factor contribute to the unsafe conditions or unsafe acts seen in this occurrence?) **Or** **Describe the risk** (***How*** does this factor increase the likelihood of human error or reduce the organization's ability to manage safety?)
Conclusion: Briefly describe what analysis will say about this issue (it may be helpful to sketch out a draft finding.	

Transportation Safety Board of Canada Bureau de la sécurité des transports du Canada

23

Safety Culture Assessment Tool

Safety Culture Assessment Tool Worksheet			
Completed by:		**Date:**	
		Safety Culture Spectrum	
	Discouraging effective safety management	⟷	***Encouraging effective safety management***
	Poor	***Bureaucratic***	***Positive***
Hazard information is	Suppressed	Ignored	Actively sought
Examples	•		
Safety messengers are	Discouraged or punished	Tolerated	Trained and encouraged
Examples	•		
Responsibility for safety is	Avoided	Fragmented	Shared
Examples	•		
Failures lead to	Cover-ups	Local Fixes	Inquiries and systemic reform
Examples	•		
New ideas are	Crushed	Seen as problems	Welcomed
Examples	•		
Conclusion: Describe the extent to which the organization's safety culture is supportive of effective safety management			
1.			
Implication – Is the organization sufficiently mature to implement effective safety management practices and is the degree of oversight being provided appropriate?			

Note: Remember to update the ISIM safety analysis with any changes or additions.

Transportation Safety Board of Canada Bureau de la sécurité des transports du Canada

24

QA Tool

- Compelling argument for change:
  - Demonstrate the link
  - Apply a reasonableness test
  - Apply a sufficiency test
  - Look for generalizability
- Conclusion → Next Steps

Transportation Safety Board of Canada Bureau de la sécurité des transports du Canada

25

Benefits

- Ensure adequate data is collected to support the safety issues;
- Guide the analysis to ensure a clear link between the factual information and safety management or safety culture;
- Serve to document the analysis making the thinking accessible for the purposes of report review and approval; and,
- Provide a "standard of evidence" for demonstrating organizational and management factors and a means for self-assessment against this standard

Transportation Safety Board of Canada Bureau de la sécurité des transports du Canada

26

Contact

- Please contact me for more information or to share feedback on the use of the tools:

  joel.morley@bst-tsb.gc.ca

  (819) 953-1634

Canada

Mike Iden,
Union Pacific Railroad

Mike is the General Director Car & Locomotive Engineering at Union Pacific Railroad. He was previously employed by the Southern Railway, Electro-Motive Division of General Motors, and the Chicago & North Western, and has been a member of the Association since 1975 or 1976.

He is the longest-serving member (and chairman for the 2nd time) of the AAR Locomotive Committee (which sets AAR Standards for locomotives), and was the first chairman of the AAR Coupling Systems and Truck Castings Committee when it was formed in 2004. In October 2012 he was asked to chair the AAR's Technical Advisory Group on natural gas fuel tenders (which is this afternoon's topic.).

Mike has a bachelor of science degree in mechanical engineering from the Milwaukee School of Engineering and a Master of Management degree from Northwestern University. He is a professional engineer in several states and has an FRA certificate.

Association of American Railroads Natural Gas Fuel Tender Technical Advisory Group (AAR NGFT TAG) update

Michael E. Iden, PE, Union Pacific Railroad, for
ASSOCIATION OF AMERICAN RAILROADS
NATURAL GAS FUEL TENDER TECHNICAL ADVISORY GROUP

1

"2nd time" for natural gas as locomotive fuel

- Major R&D programs in 1990's to commercialize LNG as loco. fuel
  - Burlington Northern (BN) with Energy Conversions, Inc.
  - Union Pacific (UP) with GM-EMD and GE Transportation
  - Gasrail USA (included UP and GM-EMD)
  - None ever reached commercialization for various reasons
  - Arrival of US EPA locomotive emissions standards effective 1-1-2000

- Big difference this time: huge discoveries of natural gas

Michael E. Iden, PE, Union Pacific Railroad, for
ASSOCIATION OF AMERICAN RAILROADS
NATURAL GAS FUEL TENDER TECHNICAL ADVISORY GROUP

2

How could natural gas affect emissions?

- Improvements made in locomotive emissions using diesel fuel
  - EPA Tiers 0/0+, 1,1+, 2/2+, 3 and soon Tier 4 (and large capex acquisitions!)
  - Criteria emittants NO_x, PM, HC, CO
- Improvement in air quality nationwide since CAA & amendments
- But, continuing/growing challenges meeting NAAQS in some areas
- SIPs (state implementation plans to meet US EPA targets)
- Global emphasis on GHG

Michael E. Iden, PE, Union Pacific Railroad, for ASSOCIATION OF AMERICAN RAILROADS NATURAL GAS FUEL TENDER TECHNICAL ADVISORY GROUP

3

Natural gas & diesel fuels compared

- Natural gas (from the ground) is a mixture of hydrocarbons
  - Methane (~95%) CH_4
  - Ethane C_2H_6
  - Propane C_3H_8
  - Butane C_4H_{10}
  - All are gases at "normal" atmospheric temperature and pressure
- Diesel (as distilled from petroleum) is a mixture of hydrocarbons
  - Typically $C_{10}H_{20}$ through $C_{15}H_{28}$
  - Always a liquid at "normal" atmospheric temperature and pressure

Michael E. Iden, PE, Union Pacific Railroad, for ASSOCIATION OF AMERICAN RAILROADS NATURAL GAS FUEL TENDER TECHNICAL ADVISORY GROUP

4

How are LNG and CNG produced?

- LNG is liquefied natural gas
  - Natural gas (methane, propane, ethane, butane) is refrigerated ("chilled")
  - It condenses into a liquid at a **temperature of -260°F**
  - **Volume is reduced 1/600th**
  - **Must be stored in a cryogenic tank (double-shell, vacuum insulated)**

- CNG is compressed natural gas
  - Natural gas is compressed to pressures of about 3000-to-5000 psi
  - **Volume is reduced 1/100th**
  - **Must be stored in a pressure tank**

Michael E. Iden, PE, Union Pacific Railroad, for ASSOCIATION OF AMERICAN RAILROADS NATURAL GAS FUEL TENDER TECHNICAL ADVISORY GROUP

RSI/CMA

5

US frt. RR energy: 0.5% of total 97.4 Quads

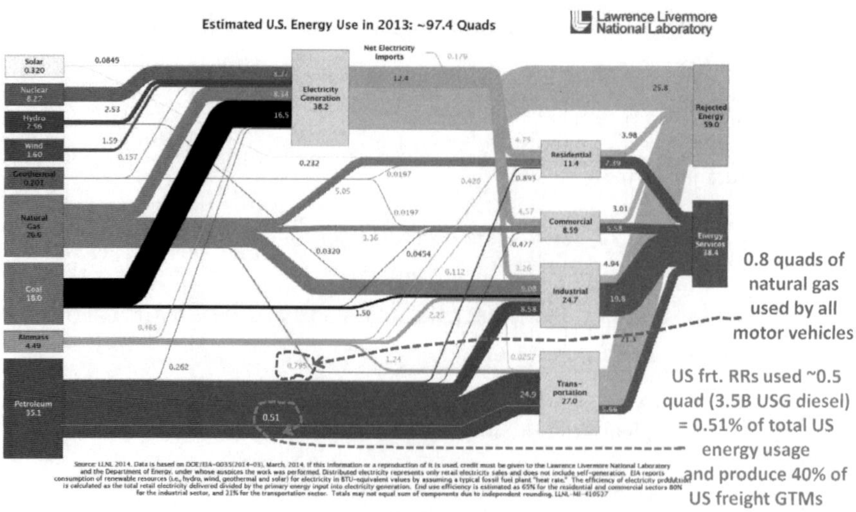

Michael E. Iden, PE, Union Pacific Railroad, for ASSOCIATION OF AMERICAN RAILROADS NATURAL GAS FUEL TENDER TECHNICAL ADVISORY GROUP

RSI/CMA

6

Canadian frt. RR energy: 0.06% of total 13.3 Quads

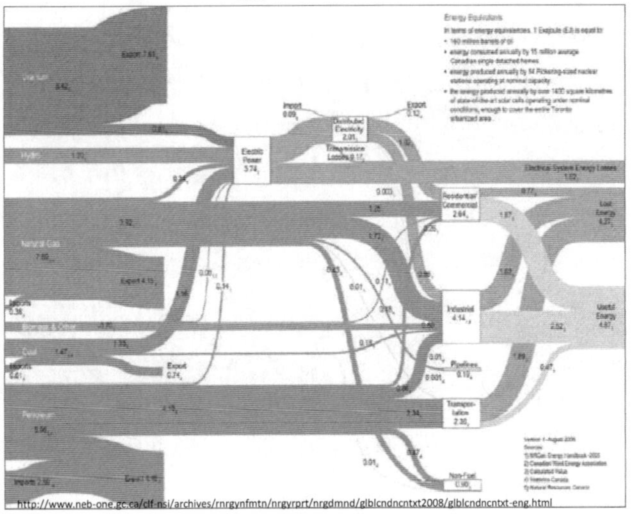

http://www.neb-one.gc.ca/clf-nsi/archives/rnrgynfmtn/nrgyrprt/nrgdmnd/glblcndncntxt2008/glblcndncntxt-eng.html

Can we convert rail to 100% natural gas?

- Yes, but unlikely
  - Gas engines don't react well to changing loads and especially altitudes
  - Dual-fuel (diesel pilot ignites natural gas) is likeliest route
- Economics, economics, economics
- Even if we did convert to 100% natural gas …

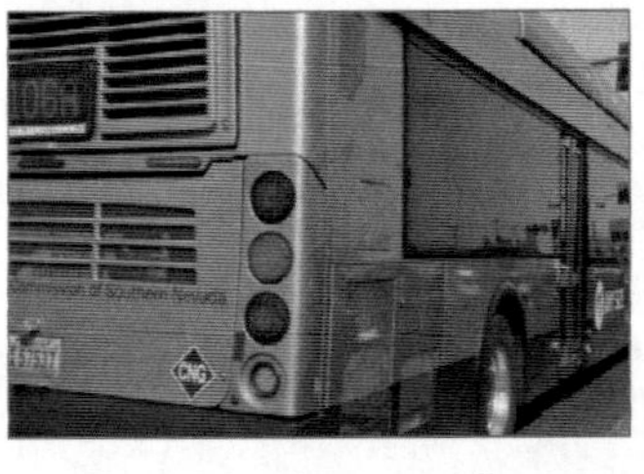

Motor vehicle use of natural gas would still exceed freight RR use

LNG as a freight RR fuel: 5 key challenges

1. Availability of natural gas (... now massive surpluses ...)
2. Liquefaction infrastructure (gas-to-LNG production)
3. Locomotives/engines which can use natural gas fuel
4. **Fuel tenders to transport/supply in-motion " " "**
5. Integration of all-above into N. American freight rail network

How do we migrate in this direction?

All 5 requirements must be met!

Michael E. Iden, PE, Union Pacific Railroad, for ASSOCIATION OF AMERICAN RAILROADS NATURAL GAS FUEL TENDER TECHNICAL ADVISORY GROUP

9

Impact of fuels & their storage on operating range

- Differences in volumetric energy density (Btu's per USG or ImpG)
  - LNG has 60% volumetric energy content of diesel, CNG only 25%
  - Now put LNG (or CNG) in "cylindrical" tanks in place of large diesel tanks ...
  - Effective energy carried onboard is reduced even more
  - Operating range is cut dramatically
- Assuming a 4300-4400 HP freight locomotive ...
  - Diesel (onboard): 1100-1400 miles (one big diesel tank)
  - LNG (onboard): 300- 400 miles (clustered cylinders)
  - CNG (onboard): ~120 miles (clustered cylinders)
- One and only one option: use a fuel tender
  - Exceptions for low-HP, short-haul and yard units

Michael E. Iden, PE, Union Pacific Railroad, for ASSOCIATION OF AMERICAN RAILROADS NATURAL GAS FUEL TENDER TECHNICAL ADVISORY GROUP

10

Locomotive fuel tank designs

1200 HP spark-ignited switcher with multiple LNG tanks (CNG tanks would be similar)

4000 HP diesel with one single diesel fuel tank

Volume lost substituting cylinders for a large tank

Cross-section view of a conventional diesel fuel tank

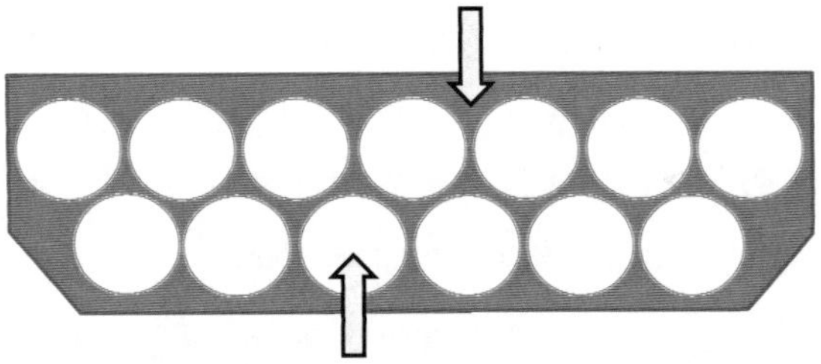

Cylinders shown can be either LNG (cryogenic double-wall tanks) or CNG (pressurized single-wall)

Tenders eliminated* in steam-to-diesel conversion

(*except for one, UP 844's tender, built in 1944)

Michael E. Iden, PE, Union Pacific Railroad, for
ASSOCIATION OF AMERICAN RAILROADS
NATURAL GAS FUEL TENDER TECHNICAL ADVISORY GROUP

13

LNG as a freight RR fuel: 5 key challenges

1. Availability of natural gas (... now massive surpluses ...)
2. Liquefaction infrastructure (gas-to-LNG production)
3. Locomotives/engines which can use natural gas fuel
4. **Fuel tenders to transport/supply in-motion " " "**
5. Integration of all-above into N. American freight rail network

How do we migrate in this direction?

All 5 requirements must be met!

Michael E. Iden, PE, Union Pacific Railroad, for
ASSOCIATION OF AMERICAN RAILROADS
NATURAL GAS FUEL TENDER TECHNICAL ADVISORY GROUP

14

What is a "fuel tender"?

- **Tender:** Locomotive Fuel Tender
  - Rolling stock designed to transport & supply fuel to a moving locomotive
  - Defined by US DOT FRA as being a "**locomotive appurtenance**"
  - Is not a "tank car" when functioning as a fuel tender
  - Long history in steam era, particularly to carry boiler water ...
  - Necessary to overcome operating range challenge

> "One of the most neglected topics of locomotive history is the tender. This indispensable auxiliary to locomotive operation was largely ignored by contemporary technical press as well. ... The earliest surviving drawing of an American locomotive tender is that of the South Carolina Railroad's "West Point" ... completed in 1831 ... The water was carried in a large cask."
>
> John H. White, Jr., A History of the American Steam Locomotive, Its Development: 1830-1880, 1968, page 233

Michael E. Iden, PE, Union Pacific Railroad, for ASSOCIATION OF AMERICAN RAILROADS NATURAL GAS FUEL TENDER TECHNICAL ADVISORY GROUP

15

Fuel tender designed for LNG fuel

- **"Tank in a tank"** (similar to DOT 113 cryogenic tank cars)
  - ***Annular space*** (between tanks) insulated, high vacuum
  - Minimizes transfer of heat from ambient environment
- Outer tank
  - Carbon steel
- **Inner tank**
  - Stainless steel, 8% nickel content, strength @ cryogenic temperatures

Michael E. Iden, PE, Union Pacific Railroad, for ASSOCIATION OF AMERICAN RAILROADS NATURAL GAS FUEL TENDER TECHNICAL ADVISORY GROUP

16

AAR effort addressing natural gas fuel tenders

- A TAG was chartered by AAR TSWC on October 4, 2012
- 1st meeting Edmonton, AB, Canada, January 29, 2013
- About 41 working days of TAG meetings since " " "

> - Develop Standard(s) for future Natural Gas Fuel Tenders for the railroad industry to support the use of natural gas (methane) as an alternative locomotive fuel
>   - Safety, crashworthiness and environmental protection
>   - Tender design and construction
>   - Tender-to-locomotive interfaces and connections
>   - Tender-to-refueling infrastructure interfaces and connections
>   - Tender interoperability and interchangeability between railroads
>   - Maintainability
> - Investigate development of dual-fuel and gas-fuel locomotives and tenders; serve as an AAR/industry interface
>   - Solicit feedback from relevant suppliers, use as TAG input
>   - Communicate and update NGFT Standard(s) process to suppliers
>
> *Mission Statement revision dated March 26, 2013*

Michael E. Iden, PE, Union Pacific Railroad, for Association of American Railroads Natural Gas Fuel Tender Technical Advisory Group

17

AAR NGFT TAG membership

*AAR rolling stock-related Committees

	Braking	Eqpmt.	HazMat	Loco.	Tank	(*other/Comm.)
AAR						XXXXXX
TTCI						XX
Amtrak				X		
BNSF	X			X	X	XXX
CN	X	X				XX
CP	X		X		X	X
CSX		X	X	X		
FEC						XX
KCS						X
NS		X		X		X
UP			X	X	X	X
FRA Safety						XXX
FRA R&D						XX

X = AAR Committee chairman, X = AAR Committee member

Michael E. Iden, PE, Union Pacific Railroad, for Association of American Railroads Natural Gas Fuel Tender Technical Advisory Group

18

What is a "TAG"?

- TAG: Technical Advisory Group
  - Does not have permanent status like an AAR Committee

- **A cool location for our 1st TAG meeting**
  - Edmonton, Alberta on January 29, 2013: **-9°F ambient, -34°F wind chill**

Michael E. Iden, PE, Union Pacific Railroad, for
ASSOCIATION OF AMERICAN RAILROADS
NATURAL GAS FUEL TENDER
TECHNICAL ADVISORY GROUP

RSI/CMA

19

Michael E. Iden, PE, Union Pacific Railroad, for
ASSOCIATION OF AMERICAN RAILROADS
NATURAL GAS FUEL TENDER
TECHNICAL ADVISORY GROUP

RSI/CMA

20

Discovery of the January sun at 53.5333° North

TAG is a joint industry & regulator activity

- FRA has been full participant since April 2013
  - Office of Safety, and Office of R&D (Rolling Stock)
  - Providing regulatory perspective
  - Allows immediate feedback of ideas
  - Consensus on accident scenarios for modeling

- Parallel crashworthiness modeling efforts
  - FRA funds engineering models by Volpe NTSC (Cambridge, MA)
  - AAR funds FEA modeling by ARA (Mountain View, CA)
  - Approaching review of all accident scenario results
  - Consensus approach to requirements in performance Standard

- Transport Canada (TC) now joining TAG

Performance Standard versus Regulation?

- AAR performance Standards
  - Insure interoperability of rolling stock between RRs
  - Standards are mandatory
  - Locomotives (and future Tenders) versus Freight Cars
    - Freight cars are governed by AAR Interchange Rules
    - Locomotives are not " " " "
    - RRs execute loco. "run through" agreements governing usage
    - RRs sign AAR Loco. Interchange Agreement guaranteeing Standardization
- Regulations
  - As with locomotive crashworthiness (S-580 & S-5506 v 49CFR229 Appendix E)
    - Likely AAR performance Standard(s) for Tenders will be incorporated by reference in some future FRA regulation(s)
- Eventually we will likely see <u>both</u> AAR Standard(s) & regulation(s)

Michael E. Iden, PE, Union Pacific Railroad, for ASSOCIATION OF AMERICAN RAILROADS NATURAL GAS FUEL TENDER TECHNICAL ADVISORY GROUP

23

NGFT TAG mission tasks$_1$: safety

- Public and employee safety
- Preventing unintended release of fuel
- Crashworthiness
  - Patterned after RSAC Locomotive Crashworthiness '98-'05
  - Sequence: accident scenario, model, analyze results & adjust/remodel
  - Incorporating modeling work from locomotives, tank cars, NGTC, etc.
  - Volpe NTSC (for FRA) & Applied Research Assoc. (for AAR)
- Regulatory & other requirements
  - FRA; AAR; NFPA; etc.
- Environmental issues
  - Minimizing/preventing fugitive emissions of CH_4 (GHG effect)

Michael E. Iden, PE, Union Pacific Railroad, for ASSOCIATION OF AMERICAN RAILROADS NATURAL GAS FUEL TENDER TECHNICAL ADVISORY GROUP

24

NGFT TAG mission tasks$_2$: multiple interfaces

- Working with locomotive OEMs & interested parties
- Working with potential tender manufacturers
- Mechanical interface (hoses, cables)
- Physical interface (pressures, flow rates, temperatures)
- Regulatory interface (safety appliances, etc)
- Infrastructure interface (tender fill connections)

Michael E. Iden, PE, Union Pacific Railroad,
for
ASSOCIATION OF AMERICAN RAILROADS
NATURAL GAS FUEL TENDER
TECHNICAL ADVISORY GROUP

NGFT TAG mission tasks$_3$: interoperability

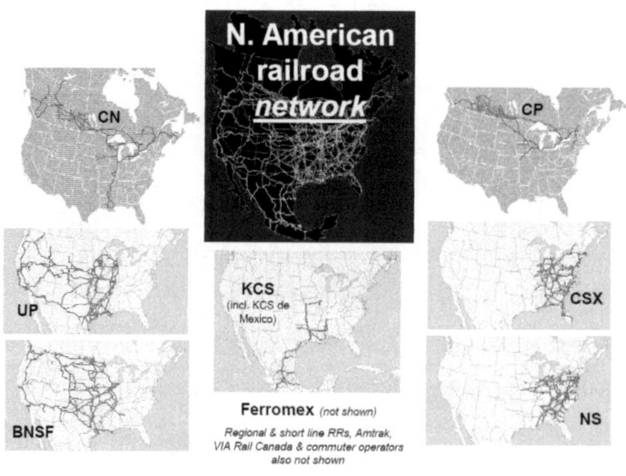

Michael E. Iden, PE, Union Pacific Railroad,
for
ASSOCIATION OF AMERICAN RAILROADS
NATURAL GAS FUEL TENDER
TECHNICAL ADVISORY GROUP

N. Am. freight RRs are *very unlike* trucking

Michael E. Iden, PE, Union Pacific Railroad, for ASSOCIATION OF AMERICAN RAILROADS NATURAL GAS FUEL TENDER TECHNICAL ADVISORY GROUP

RSI/CMA

27

NGFT TAG schedule

- The Big Question
  - "When will a NG fuel tender performance Standard be made available?"

- TAG has been working diligently to fulfill its mission
  - Roughly 6 years of "regular" AAR Committee work in <2 years
  - Unprecedented collaboration of cross-functional RR expertise
  - Locomotive, Tank Car, Equipment Engrg. Braking Systems, Hazmat
  - Advice from several working groups with others
  - 2 Town Hall & 3 supplier-type meetings held to disseminate & gather info.
  - **Oct. 21-22 Town Hall meeting in St. Louis (all interested parties)**
  - **TAG hoping to present draft performance Standard**

Michael E. Iden, PE, Union Pacific Railroad, for ASSOCIATION OF AMERICAN RAILROADS NATURAL GAS FUEL TENDER TECHNICAL ADVISORY GROUP

RSI/CMA

28

Fuel accounting (as in run thru operations!) ... ?

- LNG may introduce an entirely new set of issues:
- Most-accurate measurement of LNG is by "mass delivered"
  - Avoids measurement errors introduced by change in pressure, temperature
- Also need to measure energy content (Btu's per unit mass)
- And quality of fuel ("methane number") may be another factor
- Not quite as simple as measuring "diesel gallons"

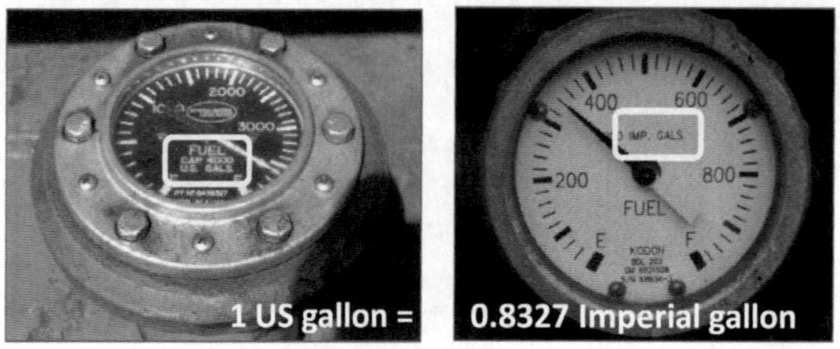

Michael E. Iden, PE, Union Pacific Railroad,
for
ASSOCIATION OF AMERICAN RAILROADS
NATURAL GAS FUEL TENDER
TECHNICAL ADVISORY GROUP

International Association of Railway Operating Officers Inc.
ESTABLISHED 1892

RSI/CMA

29

Additional resources for reading

Available from ASME (www.asme.org)

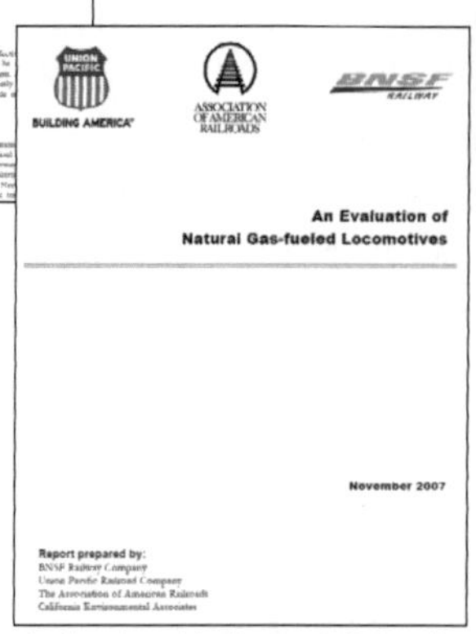

http://www.arb.ca.gov/railyard/ryagreement/112807lngqa.pdf

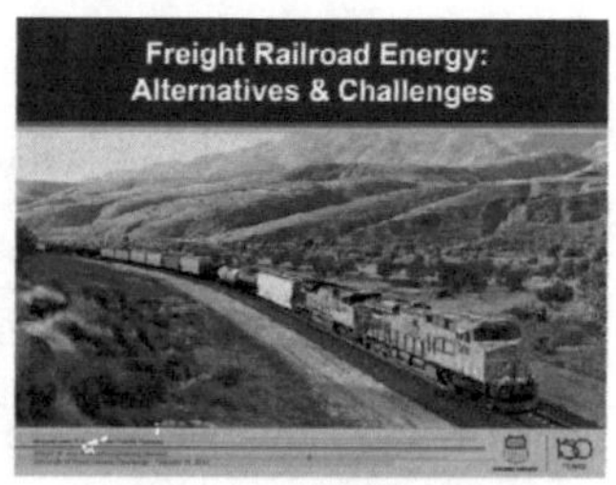

http://railtec.illinois.edu/CEE/William%20W.%20Hay/Spring2013 presentations/UP-Iden%20UIUC%20Hay%20Seminar%202013Feb15%20-%20Frt%20RR%20Energy-Alternatives%20&%20Challenges.pdf

Michael E. Iden, PE, Union Pacific Railroad,
for
ASSOCIATION OF AMERICAN RAILROADS
NATURAL GAS FUEL TENDER
TECHNICAL ADVISORY GROUP

International Association of Railway Operating Officers Inc.
ESTABLISHED 1892

RSI/CMA

30

Michael E. Iden, PE , Union Pacific Railroad,
for
ASSOCIATION OF AMERICAN RAILROADS
NATURAL GAS FUEL TENDER
TECHNICAL ADVISORY GROUP

International Association of Railway Operating Officers Inc.
ESTABLISHED 1892

RSI/CMA

31

Mark Ashbrook,
PS Technology

Mark Ashbrook received his B.A. from the University of California, Berkeley in 1997 and his M.B.A. from the University of Denver in 2008.

Mark worked as the Project Manager for Hi/Wide 3PL from 2008 to 2012. He began his current position with PS Technology in Marketing/Sales Railroad Technology.

RSI/CMA 2014

International Assoc. of Railway Operating Officers

RSI/CMA
CANADIAN RAIL SUMMIT
PALAIS DES CONGRÈS
MONTRÉAL, QUÉBEC
2014
SEPTEMBER 21-23

AGENDA

- Introduction
- PS Technology
- PST Simulation
- More than Train Handling
- PST's Vision
- Examples of PST's Vision
  - Car Inspection
  - Air Brake Tests
  - Interactive Simulated Cabs
  - Multi-User
- Simulation Roadmap

PS TECHNOLOGY

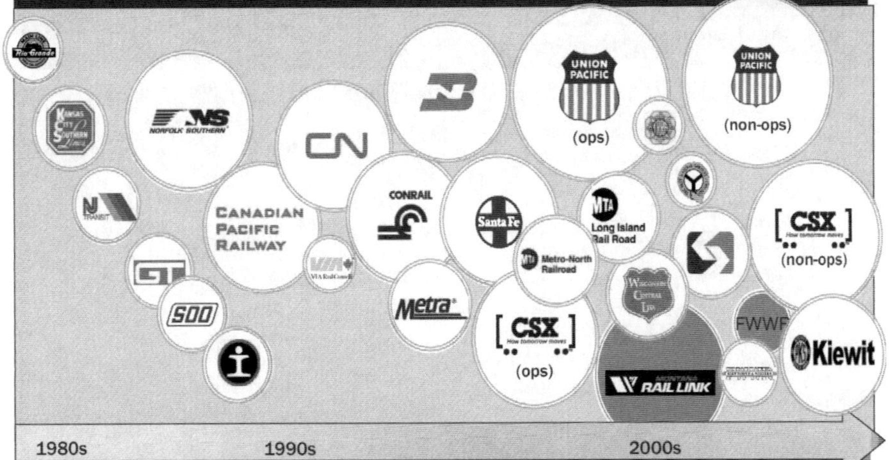

- >100 successful implementations
- all types of railroads – passenger, commuter, Class I's, shortlines
- operating crews + engineering, mechanical, and clerical workforce

PS TECHNOLOGY SIMULATION

BEYOND TRAIN HANDLING

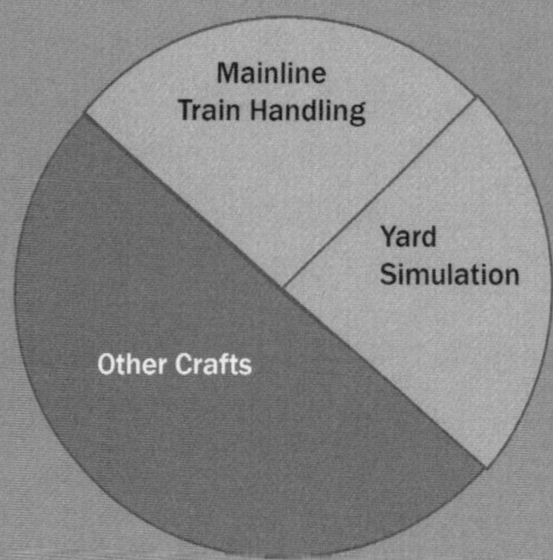

BEYOND TRAIN HANDLING

- Safety
- Efficiency
- Lower costs
- Reduce training demands on revenue generating assets
- Replicate unusual and costly occurrences

RECRUITMENT & KNOWLEDGE TRANSFER WILL BE A SERIOUS CHALLENGE

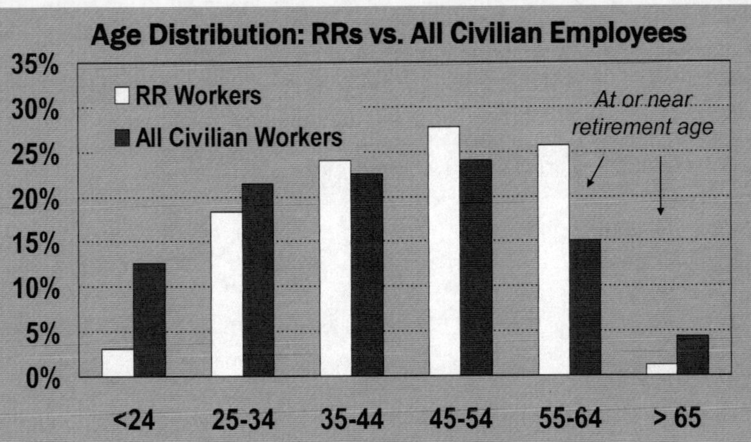

Data are for 2009. Source: Railroad Retirement Board, U.S. Census Bureau

PST SIMULATION VISION

- Training for The Entire Railroad
  - 49 CFR 240
    - Operational check rides
    - Performance skills test
  - 49 CFR 242
    - Air Brake Test
    - Hazardous Materials Training
    - Physical Characteristics Knowledge
    - Territorial Qualifications
  - 49 CFR 243
    - Track Inspection

PST SIMULATION – ONE SOLUTION

CAR INSPECTION – CAB CONTROLS

CAR INSPECTION – AIR BRAKES

DETAILED CAR FUNCTIONS

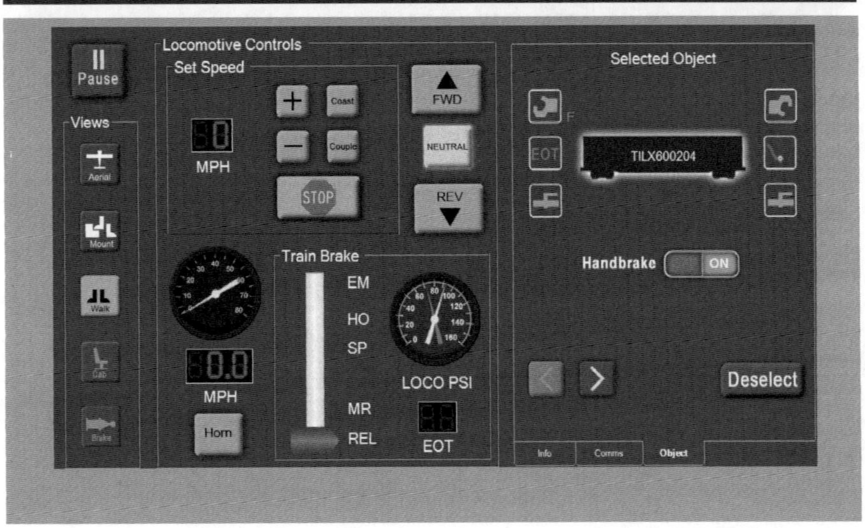

RADIO FUNCTIONALITY

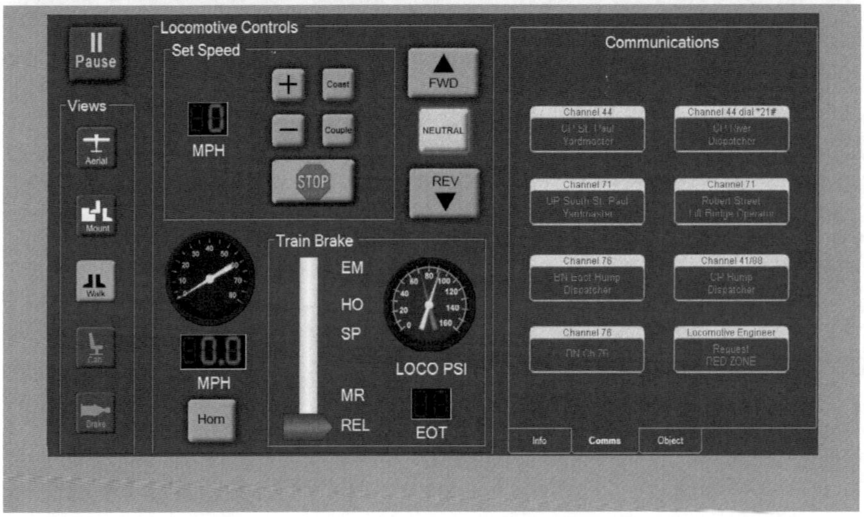

CAR INSPECTION - DEFECTS

RADIO FUNCTIONALITY

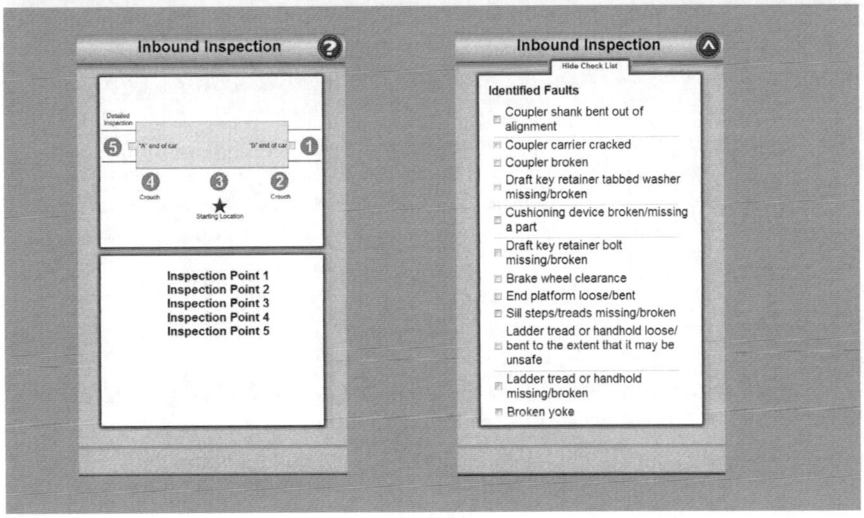

MULTI-USER

SIMULATION ROADMAP

- iClear

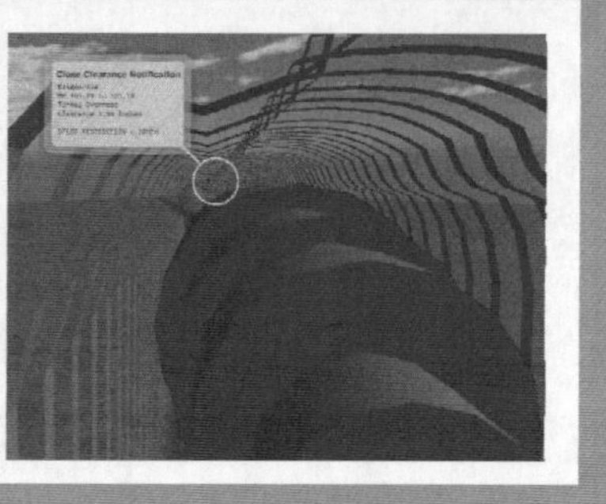

THANKS

- Questions

Jason Fox,
Union Pacific Railroad

Jason has a Master's Degree in Mechanical Engineering from Virginia Tech. He is currently Director of Mechanical Engineering for the Union Pacific Railroad. He has been with UP for eight years.

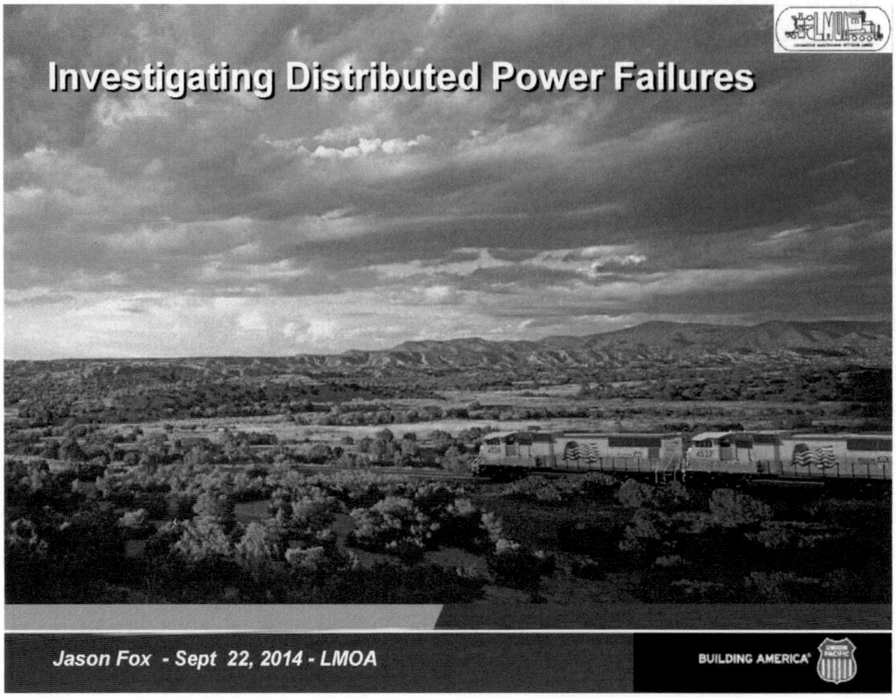

Overview of DP

- Allows crew to control a locomotive placed in a remote position on the train
- Remote unit can be placed anywhere
  - Should have a few car lengths
- Union Pacific → 3,400 DP equip
- Industry → 17,000 DP equip
- New locomotives equipped.
- Retrofit opportunities
- Longer trains and improve train handling.

2 BUILDING AMERICA

Air & Radio Links

- 450 MHz Radio link
- Brake Pipe link

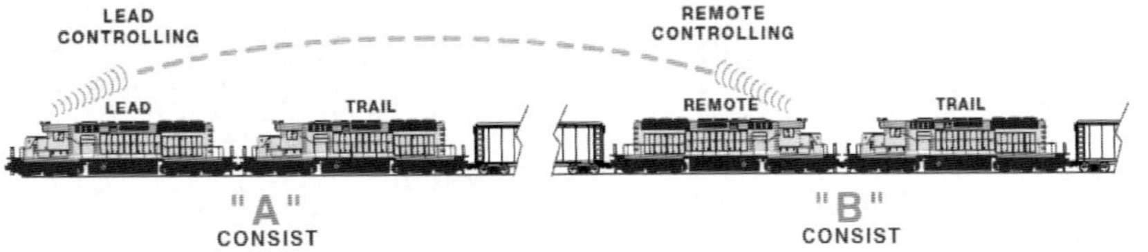

Photo: EMD, Locomotive Operation Manual Supplement for (IDP) Integrated Distributed Power. Jan 2006. P/N O00438EP

3 BUILDING AMERICA®

Communication Loss Overview

- System Designed to operate safely
- Remote will execute last command for 90 min
- Environmental and Geographical Causes
  - Tunnels
  - Mountains
  - Trees
- Message Collisions
- Hardware Failures
  - Radios
  - Cables

4 BUILDING AMERICA®

Communication Loss Idle Down

- Yellow COMM Light
  - No more Red COMM light on the UP
- DP Operation Screen freezes
- When Comm is restored
  - Remote to respond to the current requested action
  - Screen automatically updates

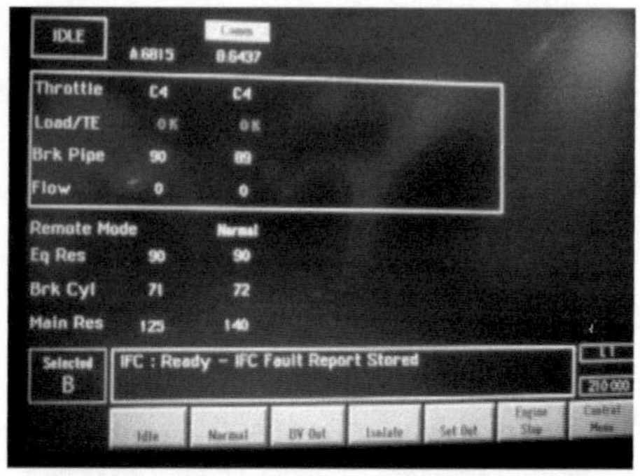

BUILDING AMERICA

Communication Loss Idle Down

- Allows crew to stop loading remote in Communication loss
  - 8 lb or greater reduction
- Remote will
  - Idle
    - Note: DB hold feature
  - Isolate
  - BV out
- Process to Recover: N.E.A.R.
  - Normalize & Execute
  - Application & Release

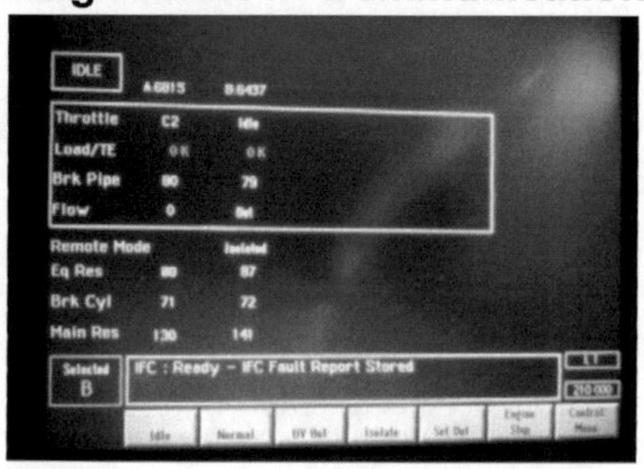

BUILDING AMERICA

The Test

BUILDING AMERICA®

Need for testing

- Crew familiarization
- Delays
- Prevention of delays
- Educational tools

8 BUILDING AMERICA®

Overview of Test

- Lead – UP6815
- Remote – UP6437
- 1x0x1 train
- 6,400 feet
- Test Leaks 20 cars from Lead and Remote

Idle Down Conditions

- Basic Communication Loss Idle Down
- Unexpected Flow
- Brake Pipe Rise
- Brake Valve Cutout in a Sustained Communication Loss

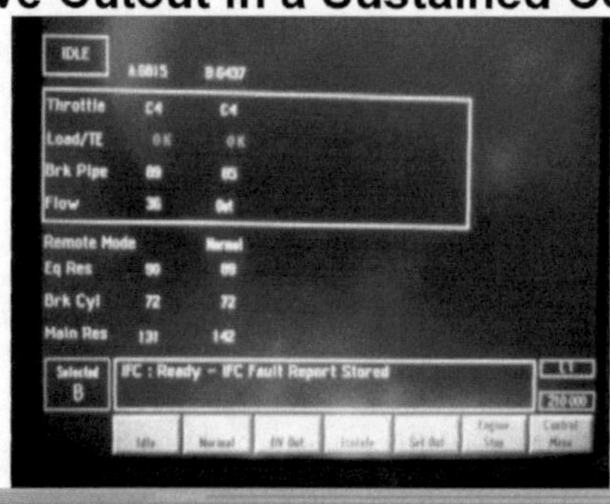

Emergency Conditions

- Low Brake Pipe
- EMV Miscompare
- TLEM – Brake pipe seperation
- Crew initiated

Emergency Conditions

- Low Brake Pipe
- EMV Miscompare
- TLEM
- Crew initiated

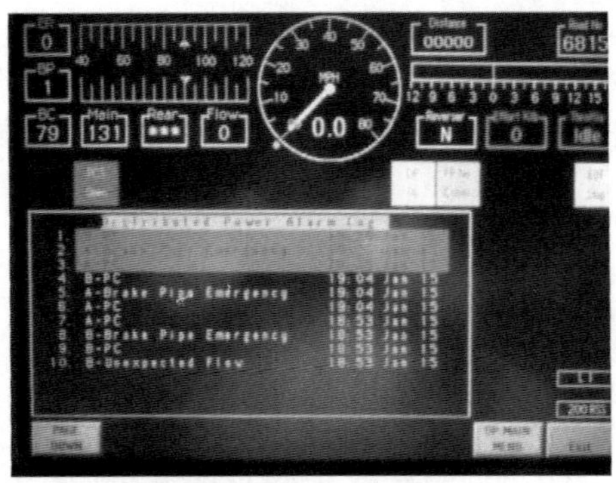

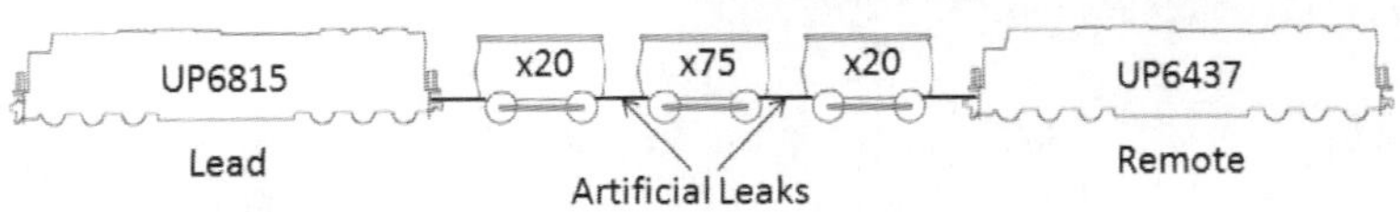

- Throttle Lockout → Brake Pipe Separation

Unlinking

- Computer resets

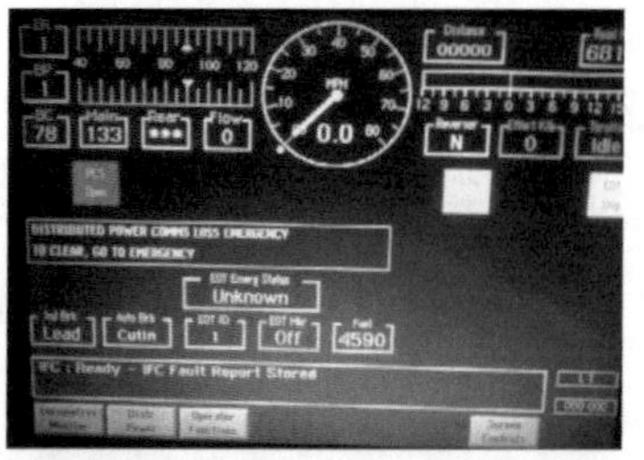

- Loss of serial communications

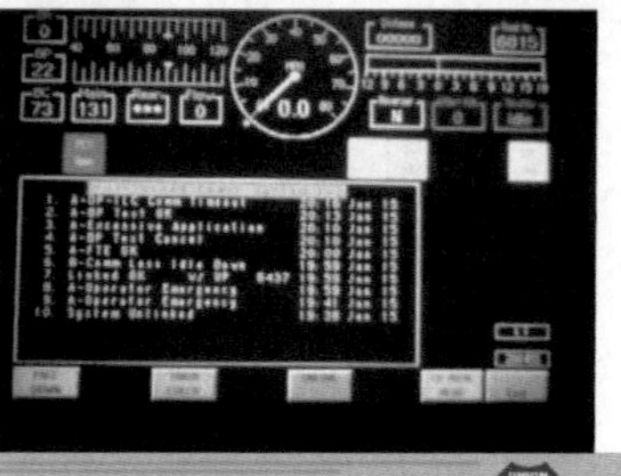

Recommendations

Clarify Messages

- Trainline Vs. Distributed Power Emergency
- Emergency from Remote XXXX
- Messages from Remotes on Lead

Eliminate Communication Loss – Expensive Option

- Add repeaters or change radio technology
- cost to the equipment and maintenance.
- Is it a wise investment?
- Heavily treed areas and flat surfaces (cliffs & buildings)
- Areas with heavy railroad traffic – message collisions
- Interoperability - manage the cutover

- System is designed to operate safely & reliably in a communication loss!

Do Not Dismiss Crew Reports

- **These men and women see this equipment every day.**
- **Ask the right questions**
- **Case Study – Unlinking and EMI mods**

Improved and Remote DP Logs

- **Must take logs from Lead & Remote**
- **Communication loss summaries**

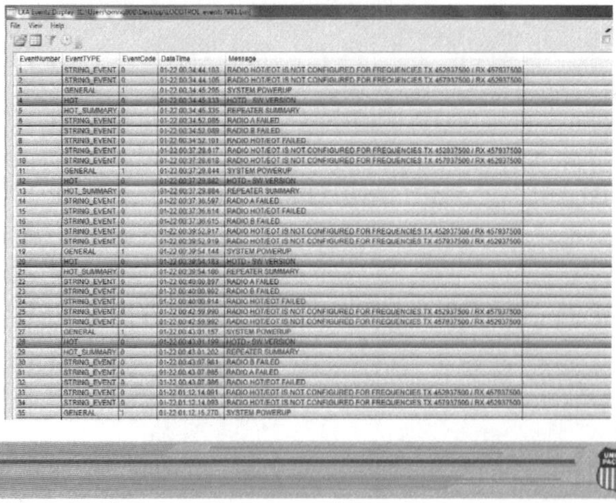

EventNumber	EventTYPE	EventCode	DataTime	Message
1	STRING_EVENT	0	01-22 00:34:44.103	RADIO HOT/EOT IS NOT CONFIGURED FOR FREQUENCIES TX 452937500 / RX 457937500
2	STRING_EVENT	0	01-22 00:34:44.105	RADIO HOT/EOT IS NOT CONFIGURED FOR FREQUENCIES TX 457937500 / RX 452937500
3	GENERAL	1	01-22 00:34:45.205	SYSTEM POWERUP
4	HOT	0	01-22 00:34:45.333	HOTD - SW VERSION
5	HOT_SUMMARY	0	01-22 00:34:45.335	REPEATER SUMMARY
6	STRING_EVENT	0	01-22 00:34:52.085	RADIO A FAILED
7	STRING_EVENT	0	01-22 00:34:52.088	RADIO B FAILED
8	STRING_EVENT	0	01-22 00:34:52.101	RADIO HOT/EOT FAILED
9	STRING_EVENT	0	01-22 00:37:28.617	RADIO HOT/EOT IS NOT CONFIGURED FOR FREQUENCIES TX 452937500 / RX 457937500
10	STRING_EVENT	0	01-22 00:37:28.618	RADIO HOT/EOT IS NOT CONFIGURED FOR FREQUENCIES TX 457937500 / RX 452937500
11	GENERAL	1	01-22 00:37:29.844	SYSTEM POWERUP
12	HOT	0	01-22 00:37:29.882	HOTD - SW VERSION
13	HOT_SUMMARY	0	01-22 00:37:29.884	REPEATER SUMMARY
14	STRING_EVENT	0	01-22 00:37:36.597	RADIO A FAILED
15	STRING_EVENT	0	01-22 00:37:36.614	RADIO HOT/EOT FAILED
16	STRING_EVENT	0	01-22 00:37:36.615	RADIO B FAILED
17	STRING_EVENT	0	01-22 00:39:52.917	RADIO HOT/EOT IS NOT CONFIGURED FOR FREQUENCIES TX 452937500 / RX 457937500
18	STRING_EVENT	0	01-22 00:39:52.919	RADIO HOT/EOT IS NOT CONFIGURED FOR FREQUENCIES TX 457937500 / RX 452937500
19	GENERAL	1	01-22 00:39:54.144	SYSTEM POWERUP
20	HOT	0	01-22 00:39:54.183	HOTD - SW VERSION
21	HOT_SUMMARY	0	01-22 00:39:54.186	REPEATER SUMMARY
22	STRING_EVENT	0	01-22 00:40:00.897	RADIO A FAILED
23	STRING_EVENT	0	01-22 00:40:00.902	RADIO B FAILED
24	STRING_EVENT	0	01-22 00:40:00.914	RADIO HOT/EOT FAILED
25	STRING_EVENT	0	01-22 00:42:59.990	RADIO HOT/EOT IS NOT CONFIGURED FOR FREQUENCIES TX 452937500 / RX 457937500
26	STRING_EVENT	0	01-22 00:42:59.992	RADIO HOT/EOT IS NOT CONFIGURED FOR FREQUENCIES TX 457937500 / RX 452937500
27	GENERAL	1	01-22 00:43:01.157	SYSTEM POWERUP
28	HOT	0	01-22 00:43:01.199	HOTD - SW VERSION
29	HOT_SUMMARY	0	01-22 00:43:01.202	REPEATER SUMMARY
30	STRING_EVENT	0	01-22 00:43:07.981	RADIO B FAILED
31	STRING_EVENT	0	01-22 00:43:07.985	RADIO A FAILED
32	STRING_EVENT	0	01-22 00:43:07.986	RADIO HOT/EOT FAILED
33	STRING_EVENT	0	01-22 01:12:14.091	RADIO HOT/EOT IS NOT CONFIGURED FOR FREQUENCIES TX 452937500 / RX 457937500
34	STRING_EVENT	0	01-22 01:12:14.093	RADIO HOT/EOT IS NOT CONFIGURED FOR FREQUENCIES TX 457937500 / RX 452937500
35	GENERAL	1	01-22 01:12:15.270	SYSTEM POWERUP

Training crews, mechanical forces and dispatchers

- Where safe, power through comm losses
- Get familiar with known comm loss areas and configure train so that it safely powers though the communication loss
- Teach what messages mean
- Provide job aids
- Same info to mech desk
- Same info to dispatchers

19 BUILDING AMERICA

John Cotter, LB Foster Rail Technology

John has been working in the global rail and transit industry for the last 15 years primarily in the development and commercialization of friction management solutions for the wheel/rail interface. He has held positions including director of research and development for Kelan Technologies as well as engineering manager for the Portec Rail UK Group.

John is currently the Engineering Site Lead (Vancouver) for the LB Foster Rail Technologies division. He holds a bachelor's degree in Chemical Engineering from the University of Toronto as well as a Master's in Business Administration from the John Molson School of Business. John is also a registered Professional Engineer as well as a certified Project Management Professional.

Implementing Top of Rail Mobile Technology in a Class 1 Environment – A Case Study

John Cotter PEng, PMP
IAROO Technical Program
September 23rd 2014

LBFoster
Rail Technologies

The Challenge with new Technology Implementation

What is desired......................

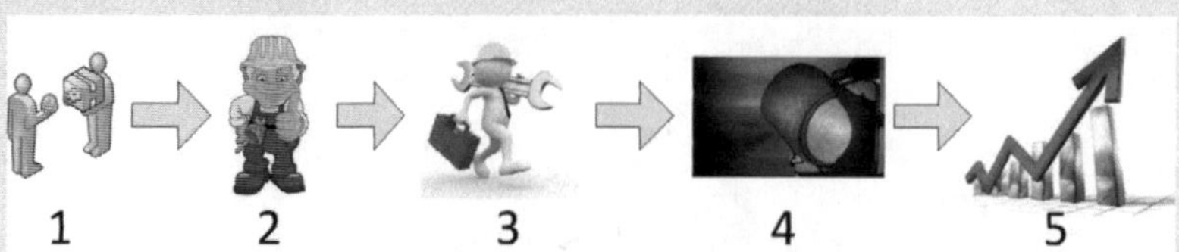

What often happens..............

LBFoster
Rail Technologies

The Specific Challenge

Trackside

Onboard

- Why do it?
  - More efficient use capital
  - More control of where to apply the friction modifier as well as how much (less product wastage)
  - Following non TOR train benefits
  - Minimizes required "on track" time (MOW)
  - Provides the same benefits (as trackside systems)

BUT FROM AN IMPLEMENTATION PERSPECTIVE.......

Trackside Systems	Onboard Systems
Significant knowledge and experience in the use of trackside applicators	Limited
Large pool of trained resources who are capable of maintaining trackside lubricators	Limited
Well established procedures for lubricator maintenance	Limited
Trackside lubricator maintenance is part of a well established maintenance routine	Not so much
Well Established reliability history	Not no much

LBFoster
Rail Technologies

OVERVIEW OF FRICTION MODIFIERS & BENEFITS OF FRICTION CONTROL

Benefits of Friction Control

Top of Rail (TOR) Friction Impacts:
- Lateral Forces / Clip Breakages
- Rail / Wheel Wear (TOR, Tread)
- Fuel Efficiency
- RCF Development
- Squeal Noise
- Flange Noise (indirect)
- Corrugations
- Hunting
- Derailment Potential (L/V, rail rollover)

LBFoster
Rail Technologies

TOR FM Benefits in the Literature

- "Implementation of Wayside Top of Rail Friction Control on North American Heavy Haul Freight Railways", *World Congress on Railway Research, 2006*
- "100% Effective Friction Management Strategy", *IHHA 2005*
- "The Effects of Top of Rail Friction Modifier on Wear and Rolling Contact Fatigue: Full Scale Rail-Wheel Test Rig Evaluation, Analysis and Modelling", *CM2006*
- "Top of Rail Friction Control: Reductions in Fuel and Greenhouse Gas Emissions" IHHA 2005
- "Freight Car-Based Top of Rail Friction Modifier Application System" IHHA 2005
- "Incorporation of Remote Monitoring Technology on a Train Mounted Top of Rail Friction Control Dispensing System Proceeding "IHHA 2007
- "Effects of Friction Control on Fuel Consumption using Train Energy Mode" TTCI Technology Digest 2007
- "A Review of Locomotive Fuel Savings with TOR-FM: Connecting theory and field results" ARM Wheel/Rail Seminar 2013

LBFoster Rail Technologies 7

Friction Targets

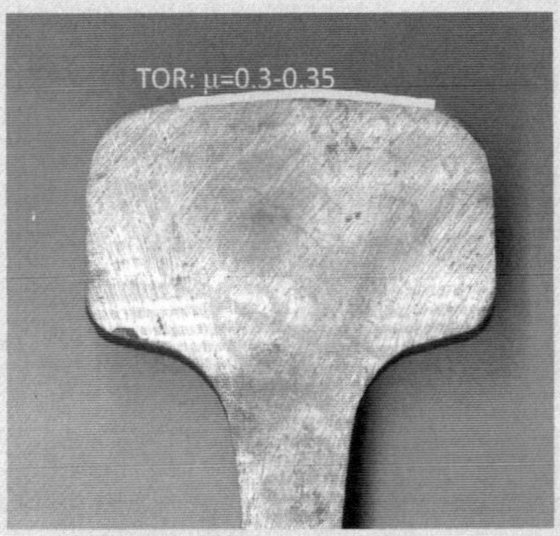

Low rail

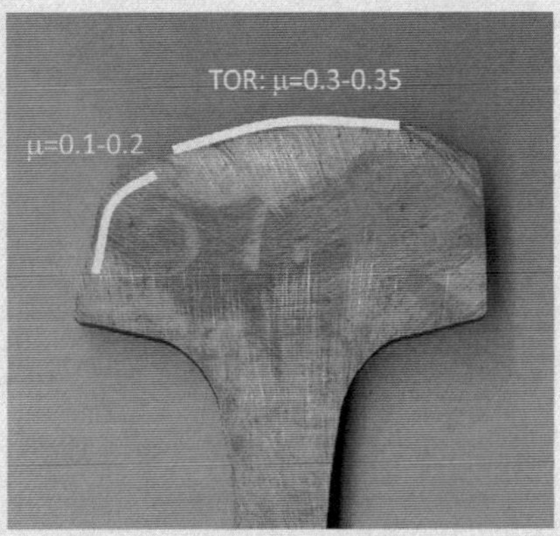

High Rail

LBFoster Rail Technologies

Third Body at Wheel/Rail Contact

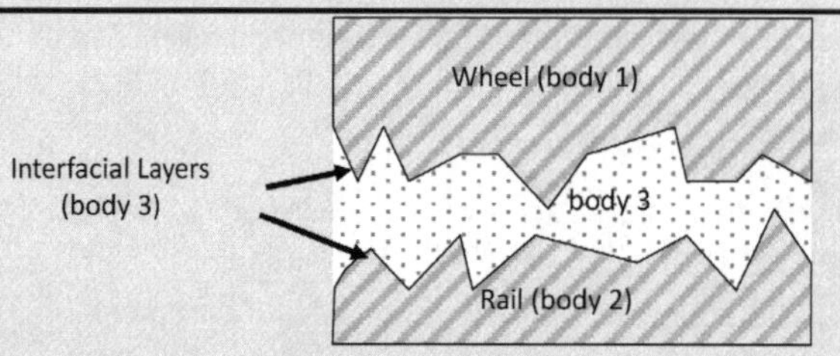

- Third Body is made up of iron oxides, sands, wet paste, leaves etc....
- Third Body separates wheel and rail surface, accommodates velocity differences and determines wheel/rail friction.
- Wheel/Rail friction depends on the shear properties / composition of the third body layer.
- Intentional addition of other materials (ie Friction Modifiers/lubricants) to this iron oxide layer can influence its' rheology.
- Impact of FM and lubricants on 3rd body rheology can differ quite significantly and have a profound impact on train handling operations.

LBFoster
Rail Technologies

Differences in CoF Behaviour
Friction Modifier versus TOR Lubricants

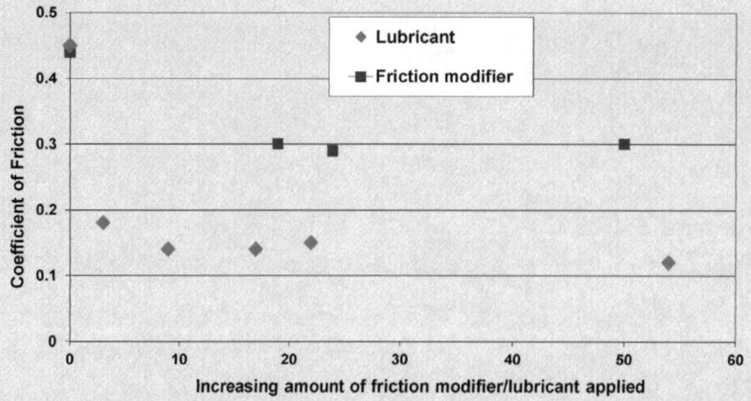

- Uniform thin film of each material applied to new rail with paint roller, quantity accurately determined and film thickness calculated from this.
- Film allowed to dry, Railhead CoF measured by a push tribometer

LBFoster
Rail Technologies

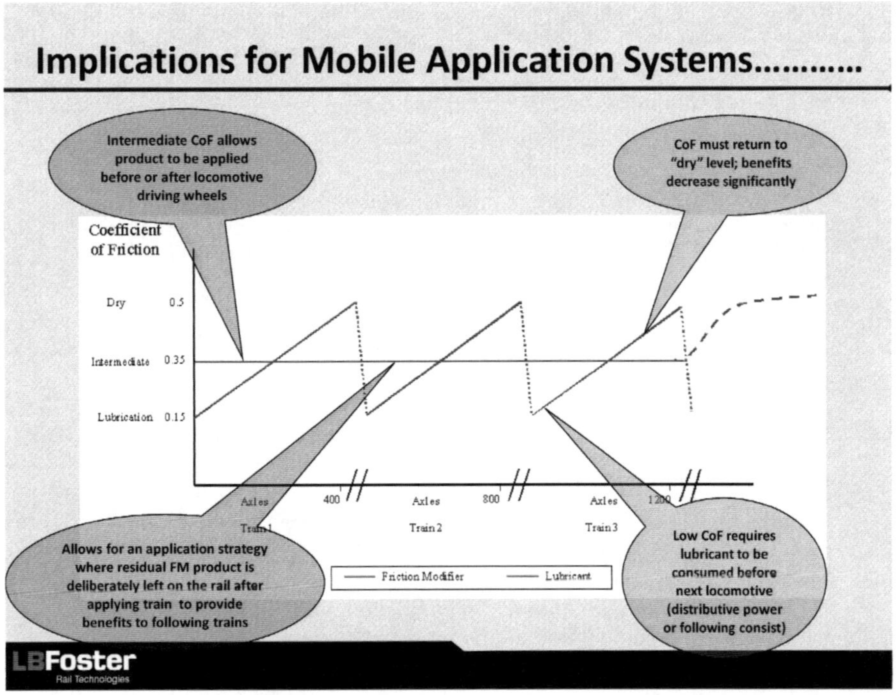
Implications for Mobile Application Systems...........
Intermediate CoF allows product to be applied before or after locomotive driving wheels
CoF must return to "dry" level; benefits decrease significantly
Coefficient of Friction
Dry 0.5
Intermediate 0.35
Lubrication 0.15
Axles
Train 1
400
Axles
Train 2
800
Axles
Train 3
1200
Allows for an application strategy where residual FM product is deliberately left on the rail after applying train to provide benefits to following trains
Friction Modifier
Lubricant
Low CoF requires lubricant to be consumed before next locomotive (distributive power or following consist)
LBFoster
Rail Technologies

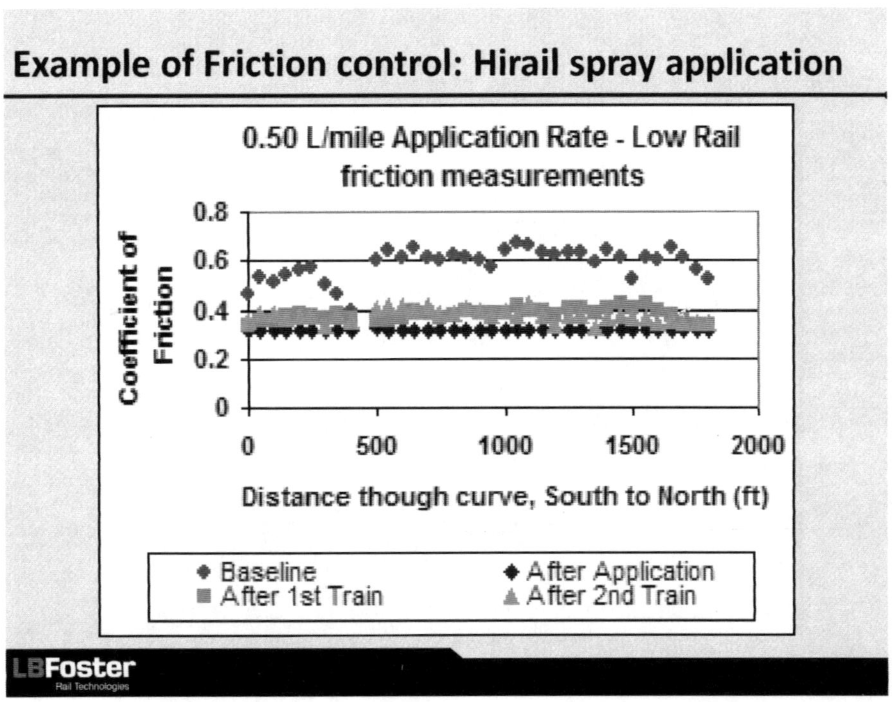
Example of Friction control: Hirail spray application
0.50 L/mile Application Rate - Low Rail friction measurements
Coefficient of Friction
0.8
0.6
0.4
0.2
0
0
500
1000
1500
2000
Distance though curve, South to North (ft)
Baseline
After Application
After 1st Train
After 2nd Train
LBFoster
Rail Technologies

GOIN' MOBILE............

Main TOR System Components

TOR outfitted Car must be first car in consist to derive air and power from locomotive

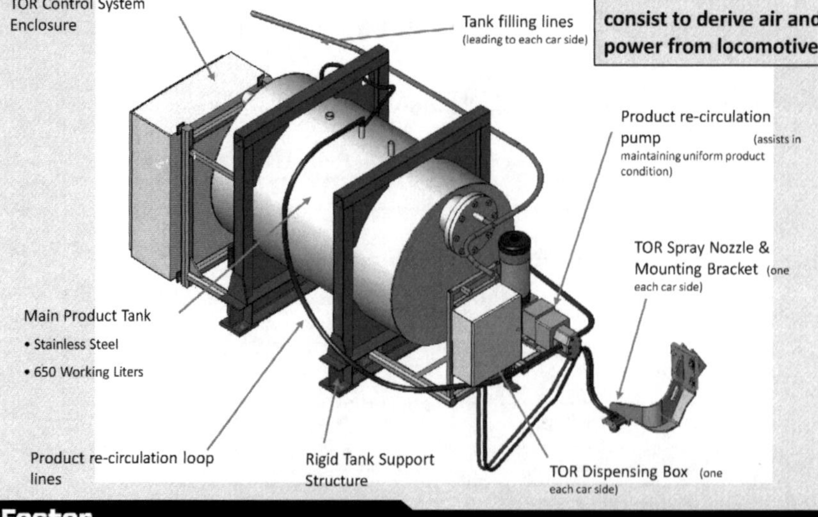

Examples of Installed Technology

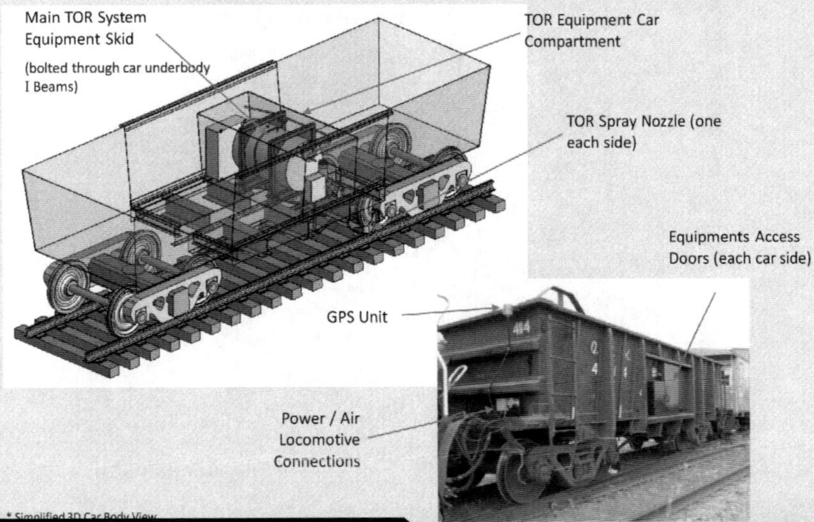

* Simplified 3D Car Body View

LBFoster
Rail Technologies

Examples of Installed Technology

LBFoster
Rail Technologies

16

Examples of Installed Technology

MU and air connections

Examples of Installed Technology

Where the Rubber Hits the Road............

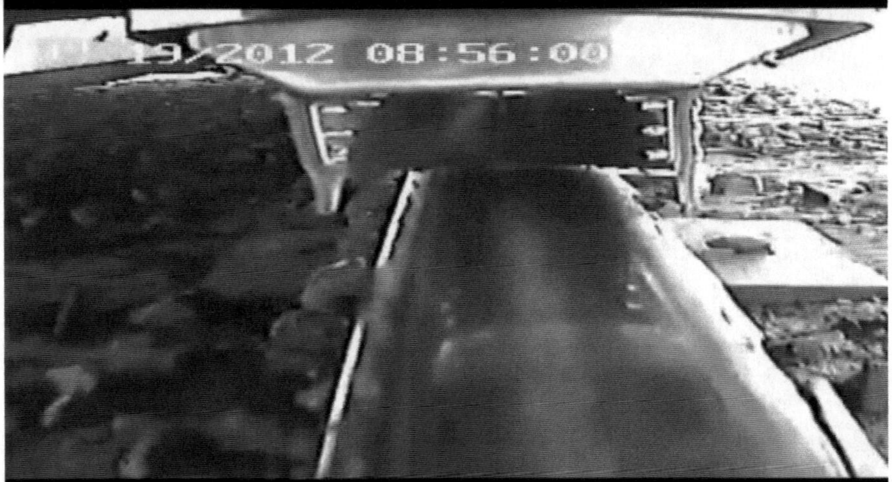

LBFoster
Rail Technologies

IT'S INSTALLED.... NOW WHAT?

Case Study: Mobile Spray System

Target

- Maintain in service fleet uptime > 90 %

Challenges

- Maintaining Air/Power Loco Connections
- **Zero Impact on Operations**
- 24/7 Maintenance Coverage
- 24/7 Tank Filling Operation
- **2 Hour Maintenance Window**
- 6 – 10 Day Turnaround Time
- Prioritizing maintenance work

LBFoster
Rail Technologies

21

Having all the Pieces in Place............

Real Time Telemetry

Trained & Equipped Technicians
(Boots on the Ground)

Maintenance Management System

Stakeholder(s) Buy In

LBFoster
Rail Technologies

22

The Need for "Real Time" Telemetry

Remote Performance Monitoring (RPM)

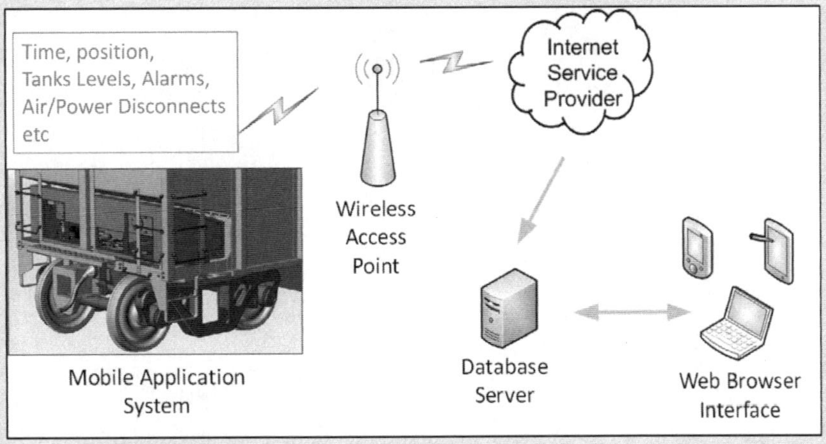

The Need for "Real Time" Telemetry

Unit Dashboard

Unit Name	Latest Record	System Status	Tank Volume (%)	Operation Status
Mobile 1	18/07/2011 06:10:33	Aux. Air Disconnected	91%	Active
Mobile 2	12/07/2011 04:01:09	Unit Not Reporting	59%	Active
Mobile 3	18/07/2011 10:03:19	Operational	62%	Active
Mobile 4	18/07/2011 09:53:59	Operational	62%	Active
Mobile 5	18/07/2011 10:21:33	Operational	88%	Active
Mobile 6	15/07/2011 15:36:22	Unit Not Reporting	94%	LB Foster Maintenance

- Can be used to schedule filling as well as focused maintenance activities

The Need for "Real Time" Telemetry: Automated Alerts

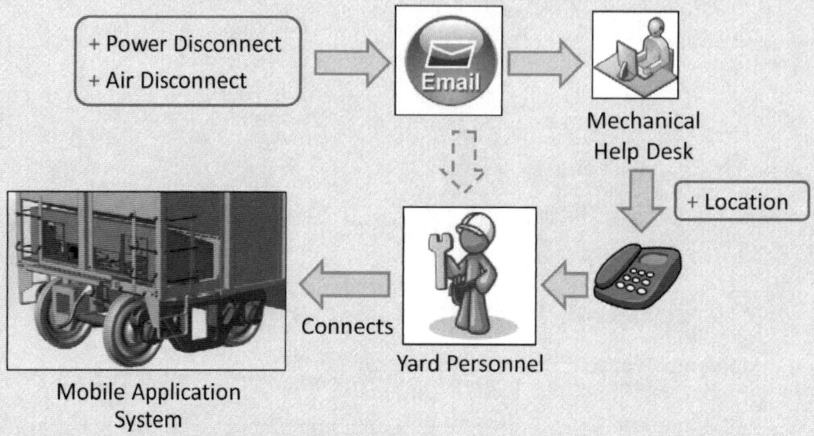

LBFoster Rail Technologies 25

Boots on the Ground.......

- On hand spares inventory
- Properly train personnel
- Design Engineering Support
- Required Tools
- "NASCAR approach" to maintenance

LBFoster Rail Technologies 26

Maintenance Management System

Records

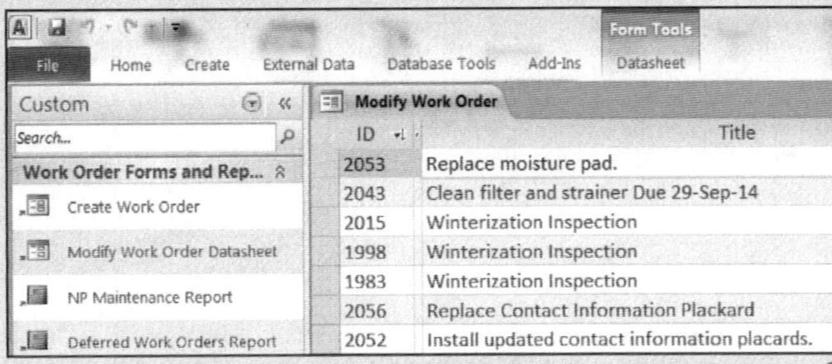

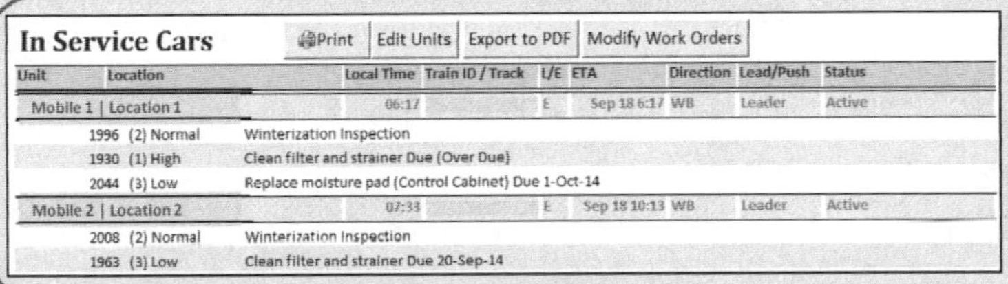

Reports

LBFoster Rail Technologies 27

Stakeholder Buy In

- Critical for successful introduction of new technology
- Maintaining (newly introduced) technology will compete for resources and attention with other high priority issues; especially those focused on throughput.
- Top down commitment/support
- Communication is key throughout organization, especially with respect to relating (new) tasks to expected benefits

LBFoster Rail Technologies 28

Uptime Achievement

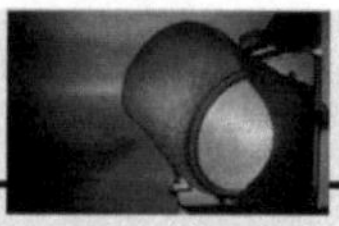

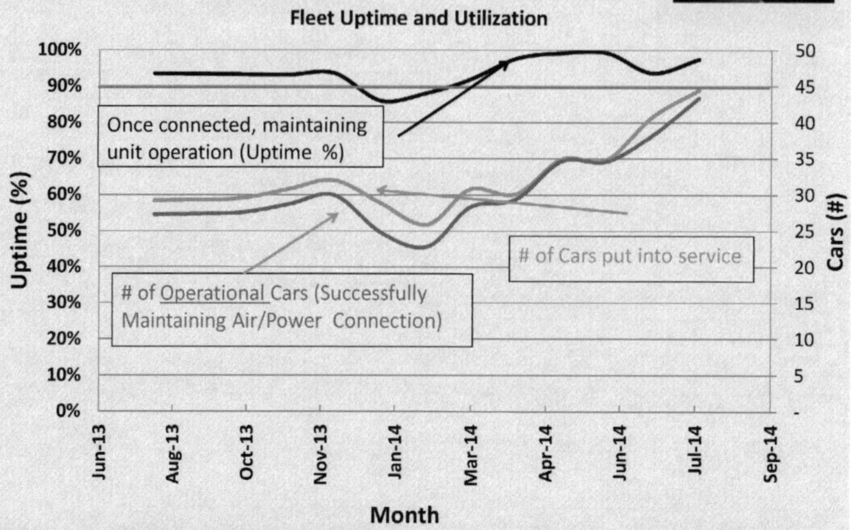

LBFoster Rail Technologies

29

Summary

- Introduction of new technology can be successful with minimal/zero impact on operations.
- Key to maximizing benefits is maintaining high uptime
- For implementation to be successful, it is suggested that the following areas be considered
  - *How to use onboard telemetry to monitor equipment performance?*
  - *How best to train and equip maintenance personnel to support the new technology?*
  - *What would a maintenance management program look like so that it doesn't impact on operations?*
  - *How do we get full commitment from all key stakeholders?*

LBFoster Rail Technologies

30

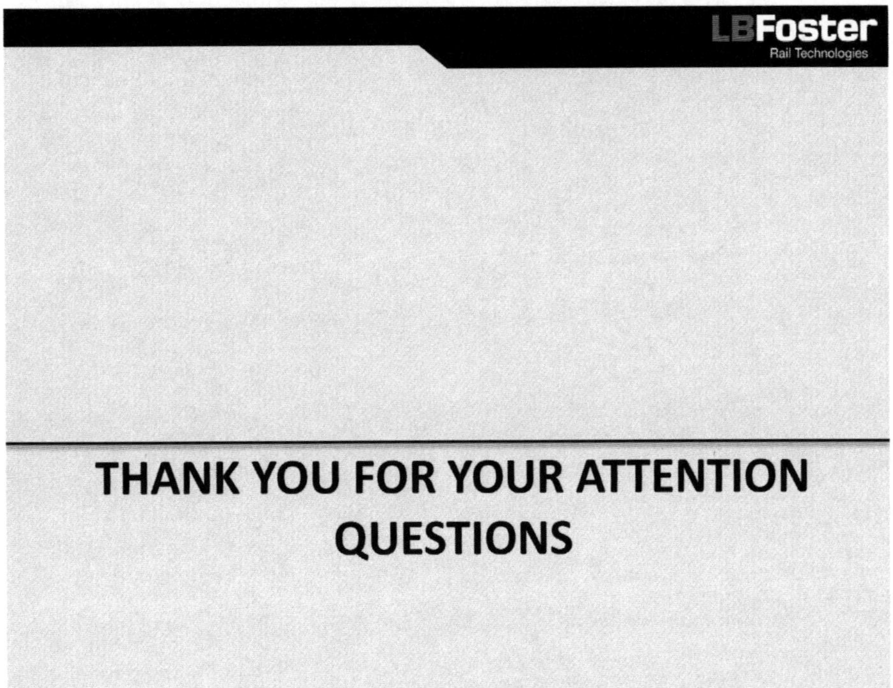
LBFoster
Rail Technologies
THANK YOU FOR YOUR ATTENTION
QUESTIONS

Dave Peltz, GE Transportation

Dave is the Chief of Global Technology for GE Transportation's Intelligent Control Solutions Division in Melbourne, Florida.

Dave's technical career began as an engineering consultant designing industrial assembly lines and automated control systems while in college.

With an Electrical Engineering degree, Dave joined the Air Force as a Systems Engineer for the MILSTAR satellite system, the military's nuclear war communications system. He invented network protocols and control algorithms for high efficiency nuclear missile re-targeting. He also designed the satellite comm distribution system for the Minute-Man and Peace-Keeper nuclear missile fields.

Dave was promoted to Program Manager of the Dual Modem upgrade program overseeing the $640 million Air Force Terminal Program source selection. Captain Peltz then went to Cape Canaveral to integrate, test, and launch satellites on the Space Shuttle and Titan rockets.

After the Air Force, Dave worked for Lockheed-Martin as Program Manager and Chief Communications System Engineer for the Eastern Launch Range, and designed the satellite comm system used for tracking or destruction of space craft launched from Florida.

Dave has been with GE Transportation for 18 years and has invented and advanced technology for products such as LOCOTROL Distributed Power, Remote Control Locomotives, the Trip Optimizer energy management system, ECP Brakes, and PTC. The in-train test equipment he has developed has significantly advanced the understanding of air brake performance and train dynamics. His brake system and pneumatics expertise and insight was instrumental in changing the long held 60 CFM departure rule to 90 CFM when using DP in Canada. Dave holds 31 U.S. and international patents in freight rail and communication

David Eldredge, GE Transportation

David was born in West Palm Beach, Florida. After attending Florida Atlantic University and graduating with a B.S degree in Mechanical Engineering in 1987, he joined Pratt & Whitney Aircraft Engines as a Mechanical Design Engineer specializing in Aeromechanical blade design. In 1998, David began his GE career in Greenville, South Carolina in the Power Systems business working in roles of increasing responsibility in compressor design technology on the F, FB, H, LMS and 6C engine platforms. In 2003, David was promoted into a Master Black Belt (MBB) position and completed his GE MBB certification in 2004. In 2005, David transitioned to GE Transportation Systems—Global Signaling in Melbourne, FL as the Quality Leader of the business and successfully led the team to ISO re-certification in 2006. In 2007, David made a career change from Engineering to Product Management of GE's Trip Optimizer Auto Train control system product line and continues to reside in Melbourne, Florida

Trip Optimizer, Energy Management and Consist Control

Presenters:
David Eldredge, GE - Product Manager
David Peltz, GE – Chief of Global Technology

Sponsor:
John Kissinger, Locomotive Fuel and Traction Consulting, Inc.

AIIRO Presentation
September 23, 2014

Imagination at work.

Trip Optimizer (TO)

Over 187,000,000 miles of service

EPA Certified

Product description

GE's Trip Optimizer is an on-board Train Control system that provides optimal fuel burn and safe train handling for locomotives. It acts similar to an aircraft's auto pilot system by automatically controlling a locomotive's throttle and dynamic brake to provide smooth operation while keeping the train on schedule and minimizing fuel use.

Customer benefits

- Sustainable fuel and emissions savings of 3-17% with 10% certified by the EPA. (variance in savings due to train weights, loco configuration, terrain properties, etc.)
- Improved train handling over a given network, by reducing variance in operator performances while utilizing automatic train handling mode.

Key features

- Closed Loop Auto Control—The train operator has the option to allow T.O. Automatic Control of Throttle and Brakes. The Operator remains under constant supervision of the train and can take charge again at anytime.
- Auto-Independent Distributed Power for auto train handling of long trains over difficult terrain with one or more remotes

Differentiating Value

- Automatic controls of throttle, dynamic braking, LOCOTROL-DP remote consists, all according to plan.
- Builds individual plan for each train over a given territory.
- Monitoring of actual HP, speed, wheel slip, engine health, wind & rolling resistance to dynamically adjust the plan.

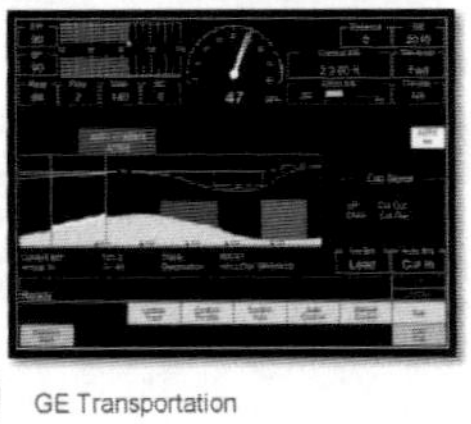

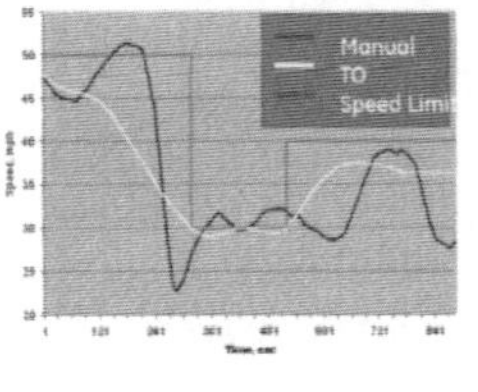

Benefits Summary

Productivity	▮▮
Lifecycle cost savings	▮▮▮
Availability/Reliability	▮▮
Fuel efficiency	▮▮▮
Safety (vs competitor)	▮▮

GE Transportation

Agenda

- Basic Trip Optimizer (TO) and how TO works
- New Features Driving Automation
  - TO SmartHPT Consist Control
  - Auto Independent Remote Control
  - Locomotive based Top of Rail Lubrication
  - Lateral Force Optimization

Trip Optimizer – How does it work?

It all starts with the Railroad's Data

- Origin-Destination
- Train Orders - Weight, Length, #Cars, #Locos
- Temporary Speed Restrictions

1

Data is then transmitted Wirelessly to the Locomotive

2

Onboard computer plans the trip; provides the plan to control system

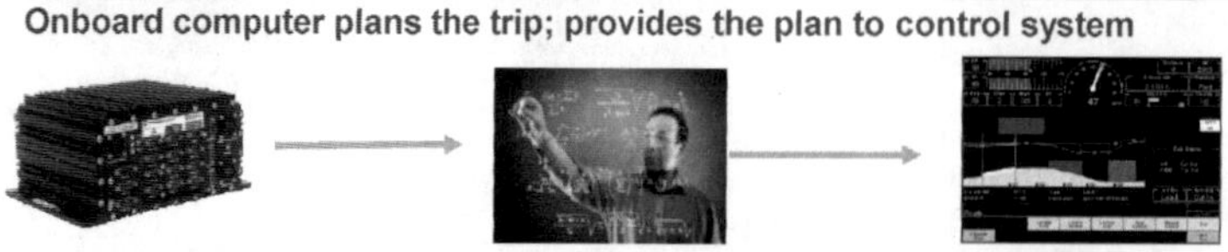

3

Locomotive microprocessors then use closed loop control to regulate the speed to the plan

4

imagination at work

4

Trip Optimizer Operation

Making all your Drivers, the Best Driver in the World!

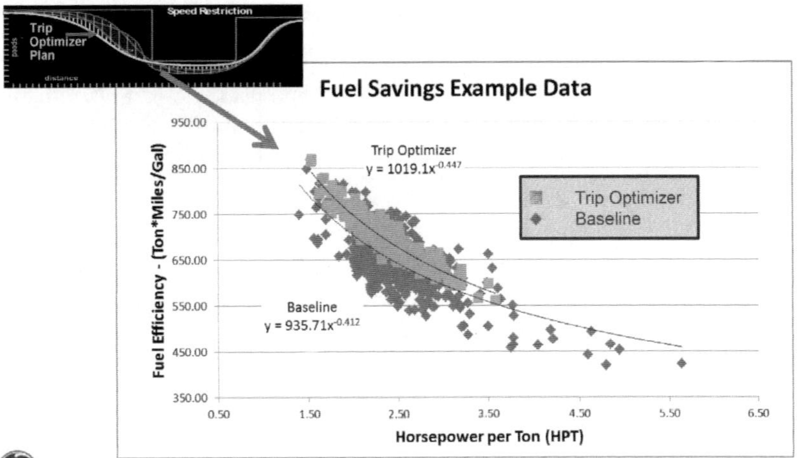

imagination at work

5

Trip Optimizer Results have been Terrific

- 3%-17% Fuel Savings
- 187,000,000 Miles Operation
- Reduced Break-in-Twos
- Reduced Over Speeds

ecomagination certified

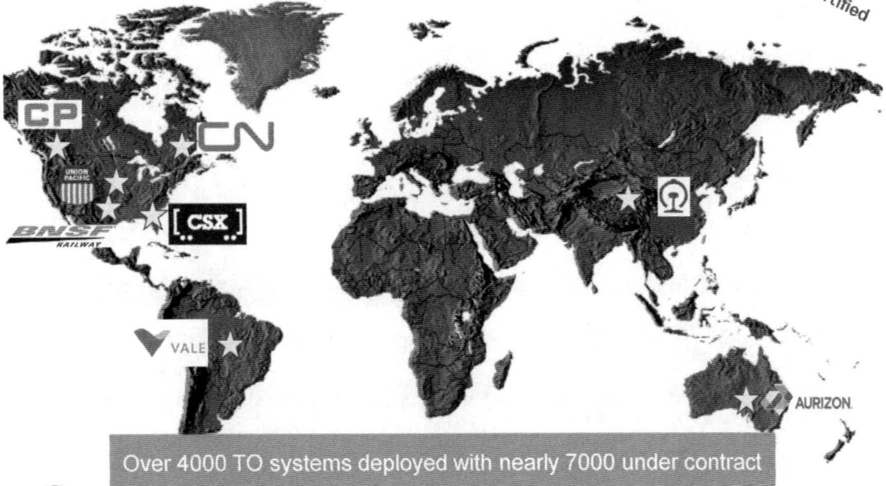

imagination at work

6
GE Transportation
6

New Trip Optimizer Features in Process...

TO Smart HPT (Horsepower Per Ton)

Trains are powered to traverse the ruling grade. TO knows route, grade & trailing tonnage to automatically (or manually) idle/isolate trail locomotives when HP not needed to maintain track speed.

- Incremental 2-5% over Trip Optimizer
- Balance consist fuel consumption, extend fueling events.
- Requires Ethernet over MU (eMU) to communicate

Testing Complete – Launch 9/29/14

TO Auto Independent Distributed Power

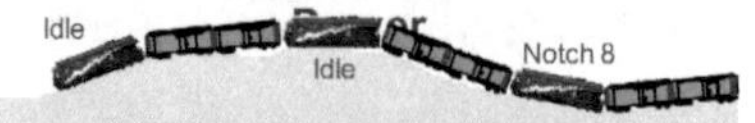

Automatic control of Locotrol Independent Mode for better Train Handling of Longer Trains over difficult Terrain

TO stays engaged in Automatic mode longer for higher fuel savings and better train handling

Successful Operation of Trains > 47,000 Tons

Launched Q4-2013

TO-PTC Integration

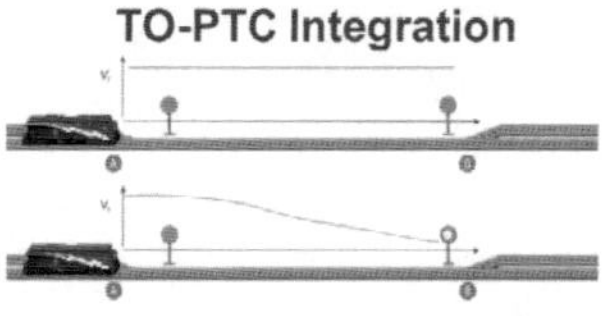

PTC passes TO new speed limits based on limits of authority. TO regulates speed to those new speed limits, system stays engaged in Auto mode longer=higher fuel savings, reduce PTC enforcements

Successful Field Test Complete

Pacing– TO Integration with GE Movement Planner

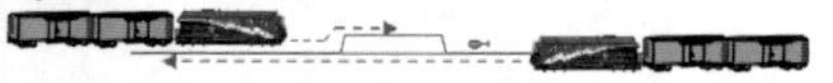

Movement Planner provides location and time of arrival, TO adjusts "Plan" and controls train to meet the MP schedule

Opposing trains meet with minimum delay and maximum fuel savings – Limit "hurry up and wait"

2-4% fuel savings with no network velocity impact

Successful Field Test Complete

imagination at work

7

Using the Trip Plan Looking Forward, SmartHPT System Does the Following:

Trail Notch Control of Lead Consist

- Idle individual trail units
- Idle ALL trail units
- Independent Consist Notching (*Future*)
- DP Remote Consist Control (*Future*)

Engine Control Switch

- ISOLATE Trail unit from RUN position
- Allows AESS to execute per specification

N8 → N8 → Idle **Example 1 – Idle unit using eMU**

N8 → Idle → Idle **Example 2 – Idle Consist using MU cable**

N6 → N6 → Isolate **Example 3 – Isolate unit using eMU**

GE Proprietary & Confidential

Trip Optimizer/ LOCOTROL® User Conference 6/22/2015

8

eMU-Ethernet over MU Cable

- Provides high speed Ethernet connectivity between lead and trail locomotives
- Connects via standard MU cable, no extra wires
- Operates through un-equipped locomotives
- Works on all locomotives that comply to "MU" standard

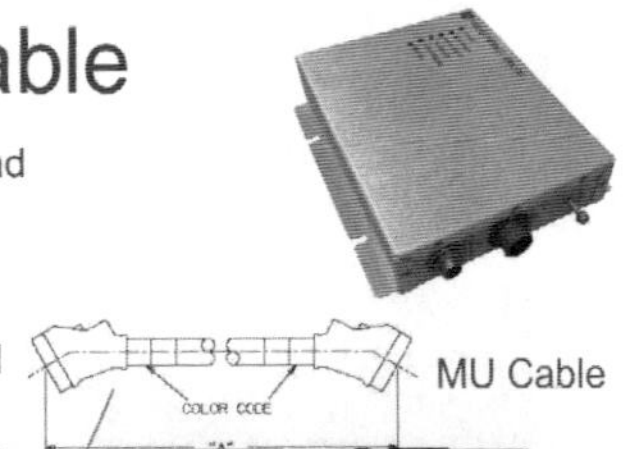

Benefits:
- 5M-40M bits/sec Ethernet bandwidth
- Network automatically established

Applications:
- Enables SmartHPT fuel option for TO
- Access Trail locomotive status at lead
- Use equipment on trail locos for redundancy (radios, PTC equipment, etc.)
- Generic Enabler for productivity applications

3,000 on order, 600 in service
Available for all locomotives today

Applications using eMU capability require OEM support

Smart HPT

2.9% Fuel Savings

3.7 % Fuel Savings

No Impact on Arrival time

Empty Coal Run Test Data

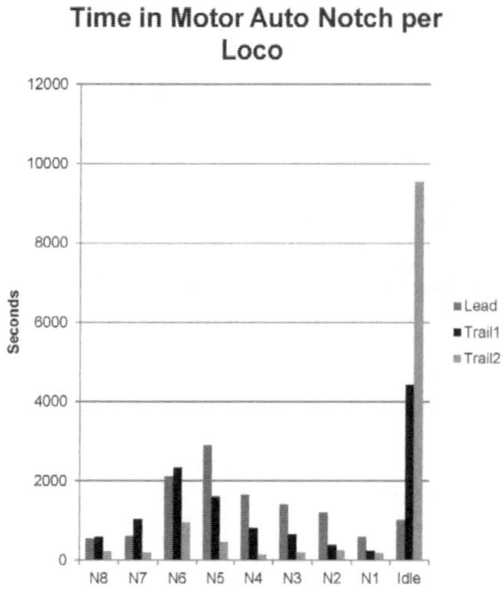

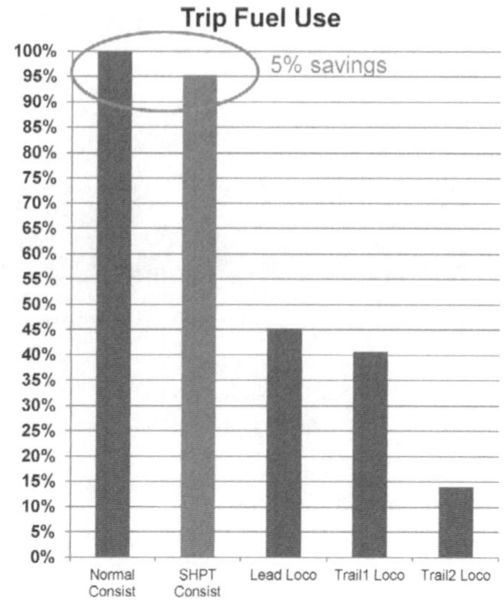

5% Fuel Savings demonstrated on a high (5) HPT Train

imagination at work

11
Trip Optimizer Enhancements
6/10/2015

TO Auto Independent DP (AIDP)

State of the Art Train Data Equipment

PPC-Daq R7
4 Channel + GPS
Pressure/0-5V selectable
100 Sample/Sec
86hr logging max (12hr/battery)
0.01 psi, 0.01", 0.01CFM

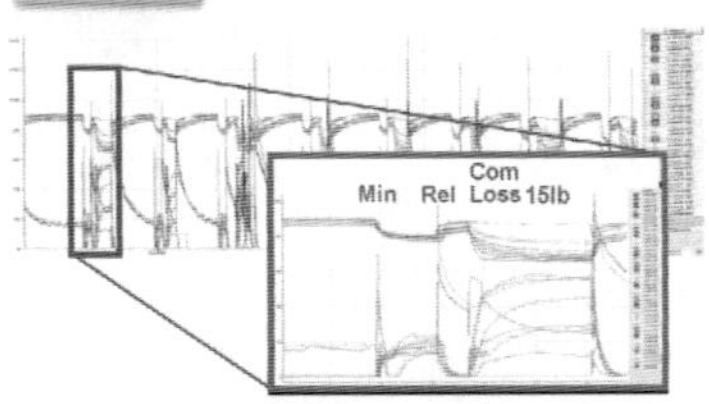

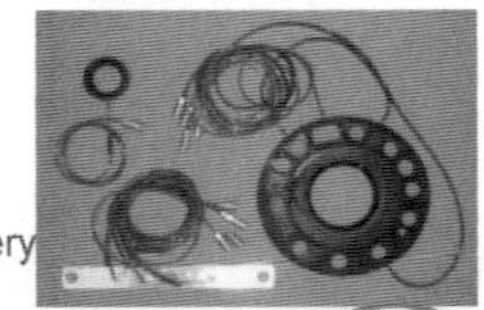

Pressure Interfaces
3/32" tubing/valve quick disconnects
- Glad-hand gasket
- Single Car Test Port
- Service Portion Gasket

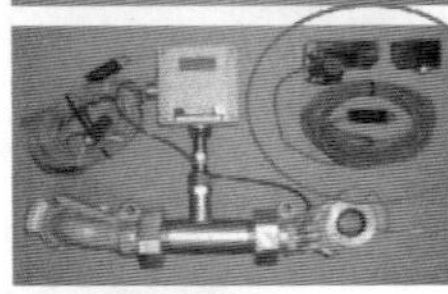

0-5V Interfaces
- 0 to 200 CFM high precision flow meter
- 36" travel string pot for slack measure

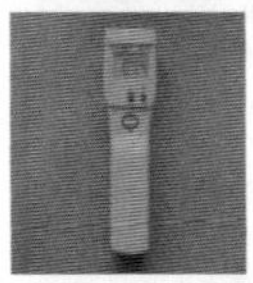

- Temp: 0.1 Degree f, 100s/s
- **Wheel Heating/Cooling Model Support**

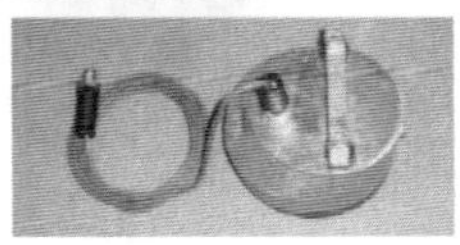

MU 27 pin connector
Converts throttle TLs to stepped voltage for capture of Notch Commands

imagination at work

13 /

Equipment Installation

imagination at work

14 /

M377-21-07 Westbound Montreal to Bellville

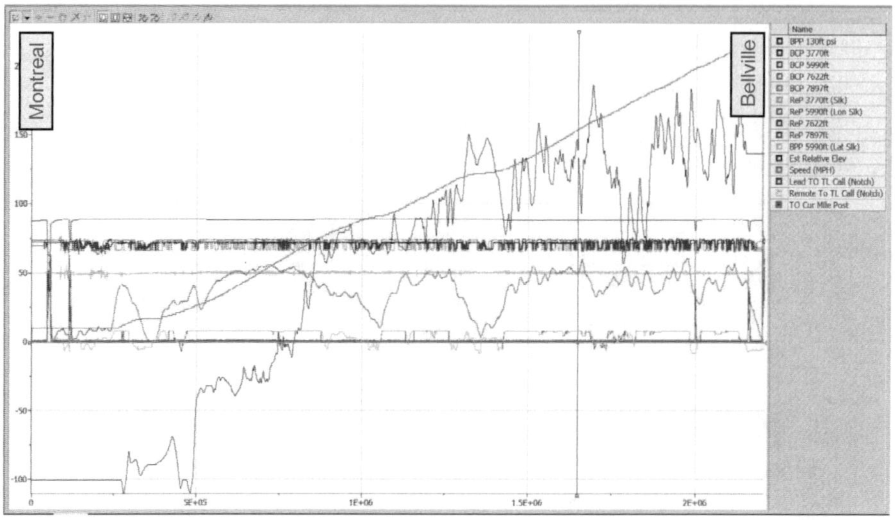

M377-21-07 Westbound Montreal to Toronto

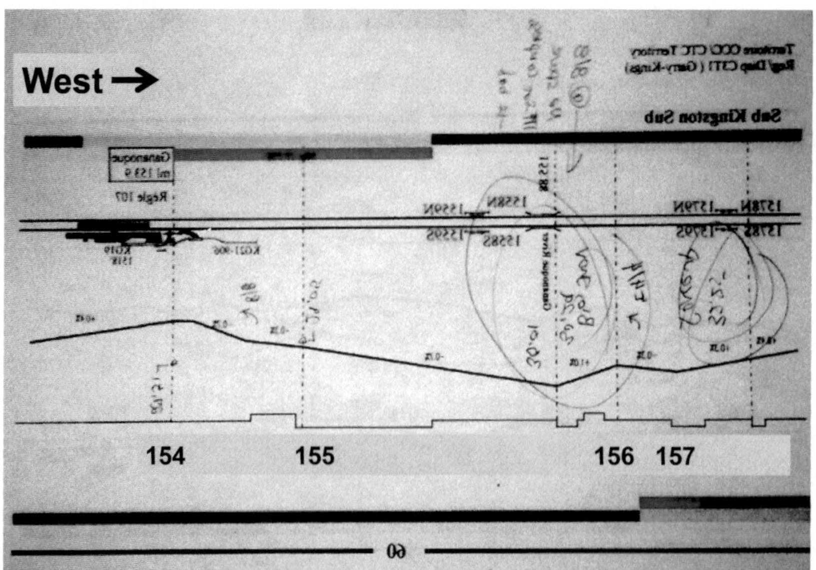

M377-21-07 Westbound Montreal to Bellville

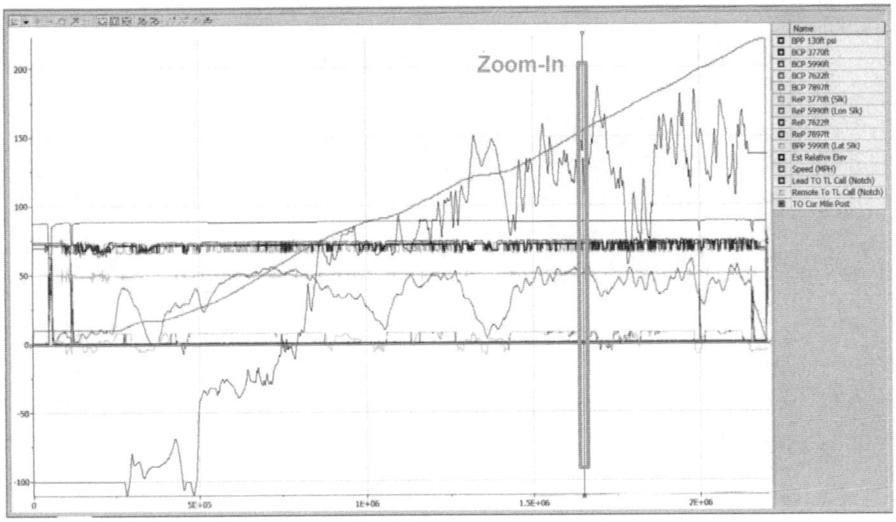

imagination at work

17 /

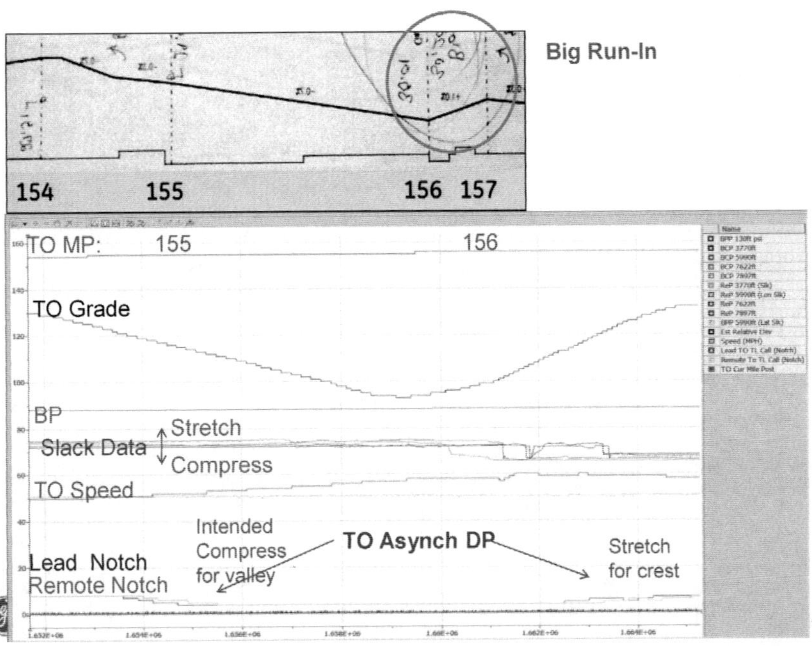

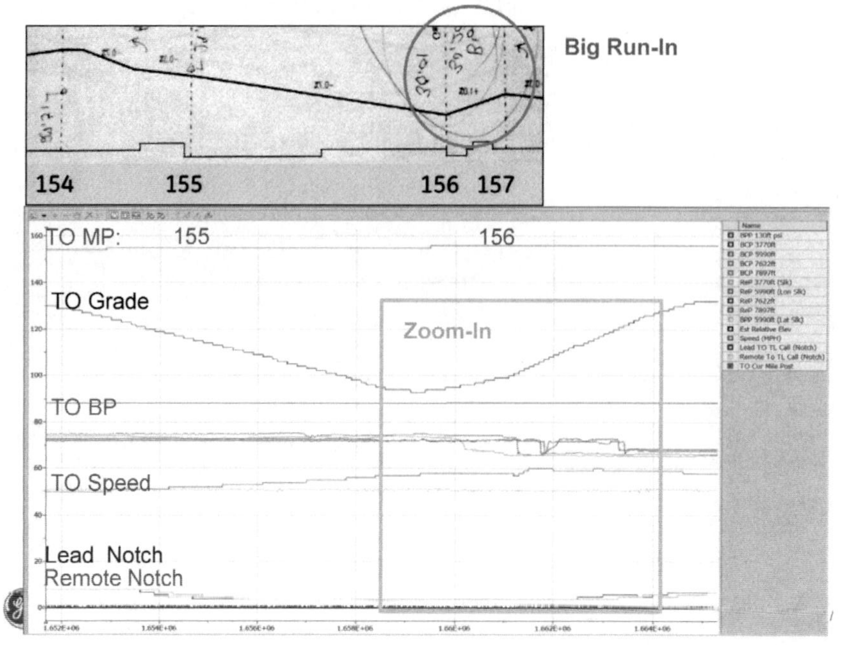
Big Run-In
154
155
156
157
TO MP:
155
156
TO Grade
Zoom-In
TO BP
TO Speed
Lead Notch
Remote Notch

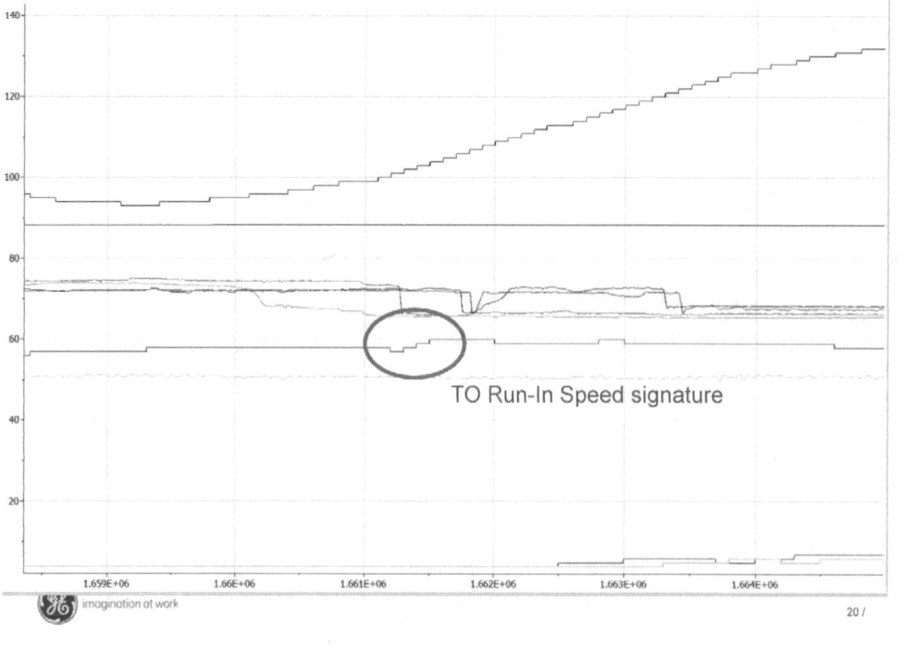
TO Run-In Speed signature
imagination at work
20 /

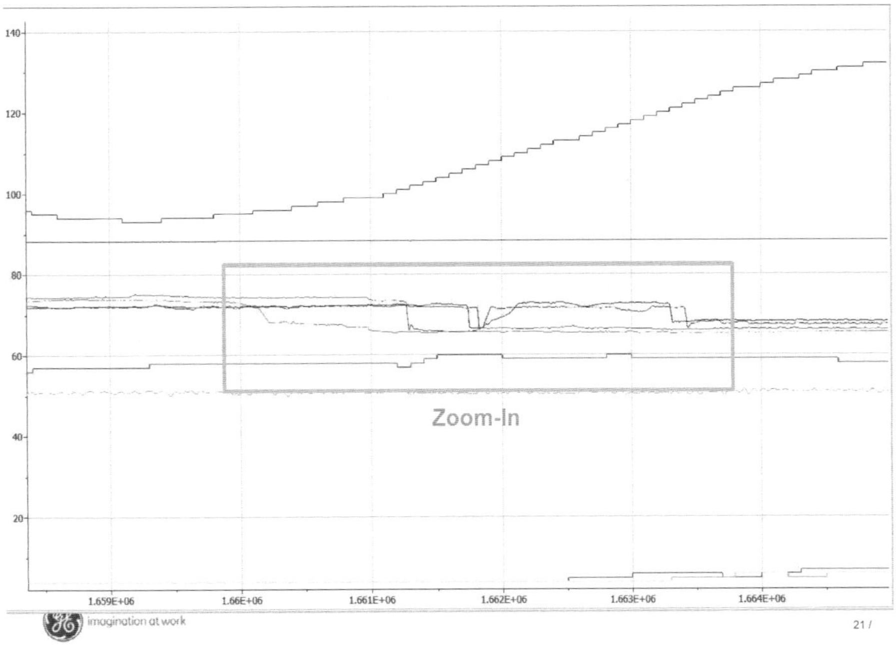

140
120
100
80
60
40
20
Zoom-In
1.659E+06
1.66E+06
1.661E+06
1.662E+06
1.663E+06
1.664E+06
imagination at work
21 /

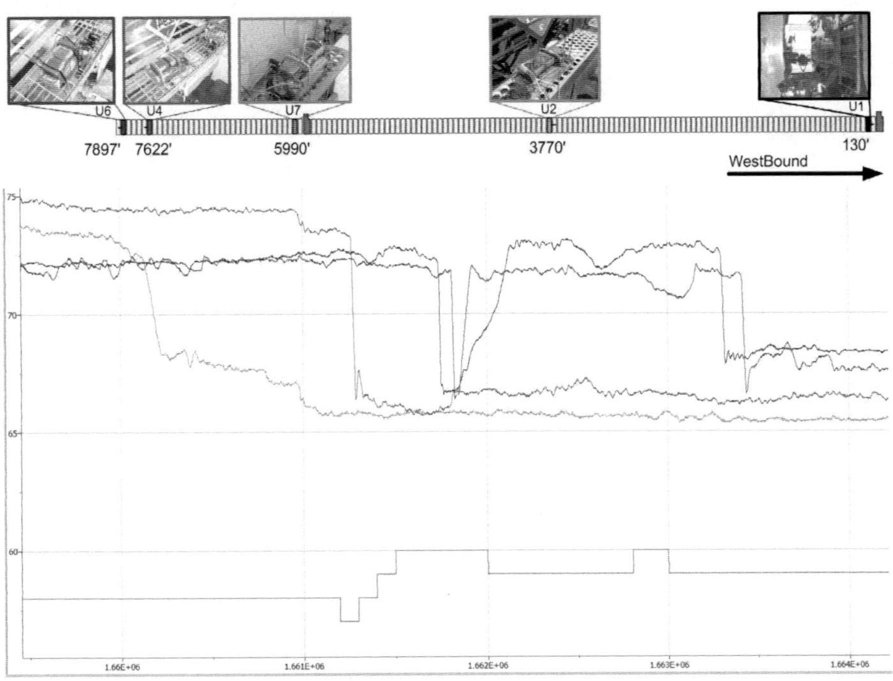

U6
U4
U7
U2
U1
7897'
7622'
5990'
3770'
130'
WestBound
75
70
65
60
1.66E+06
1.661E+06
1.662E+06
1.663E+06
1.664E+06

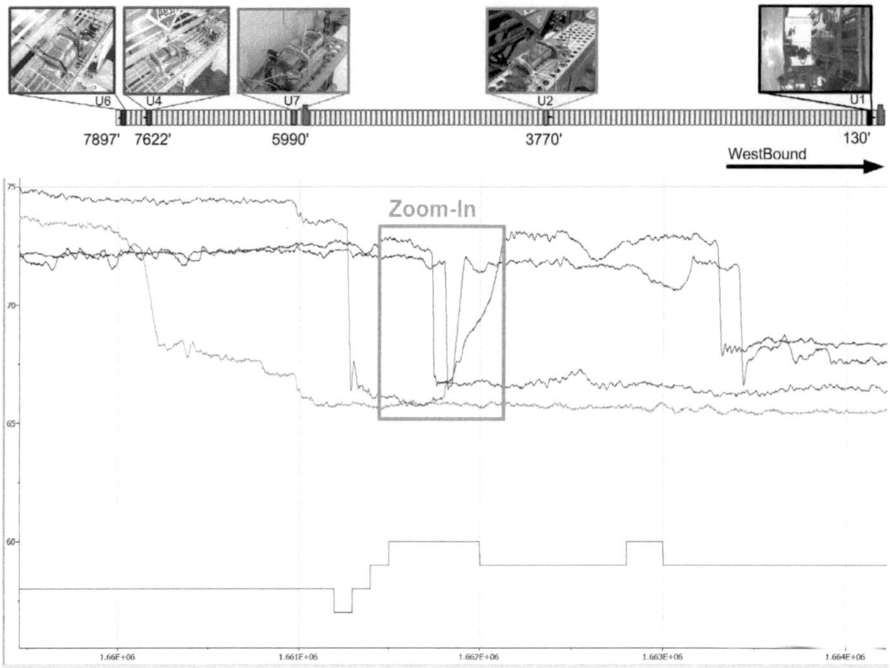

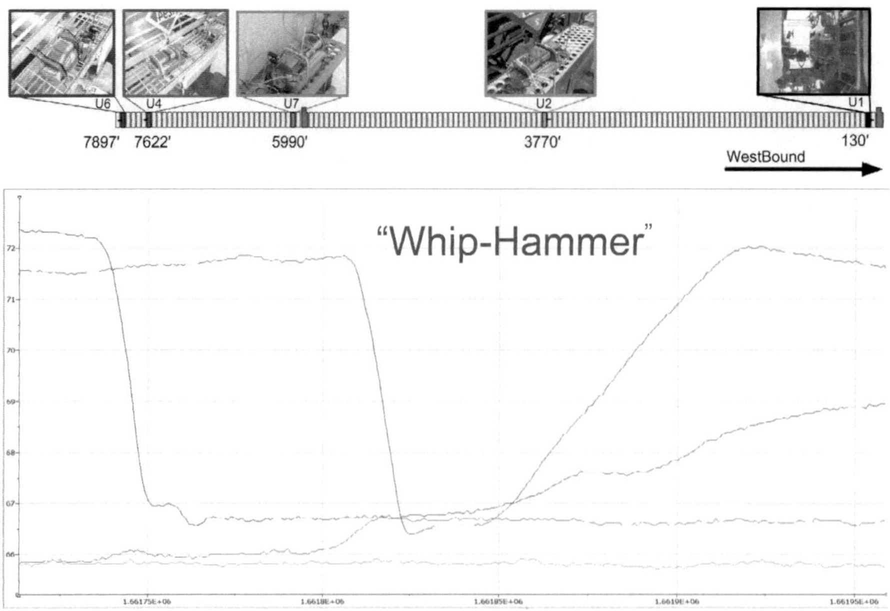

imagination at work

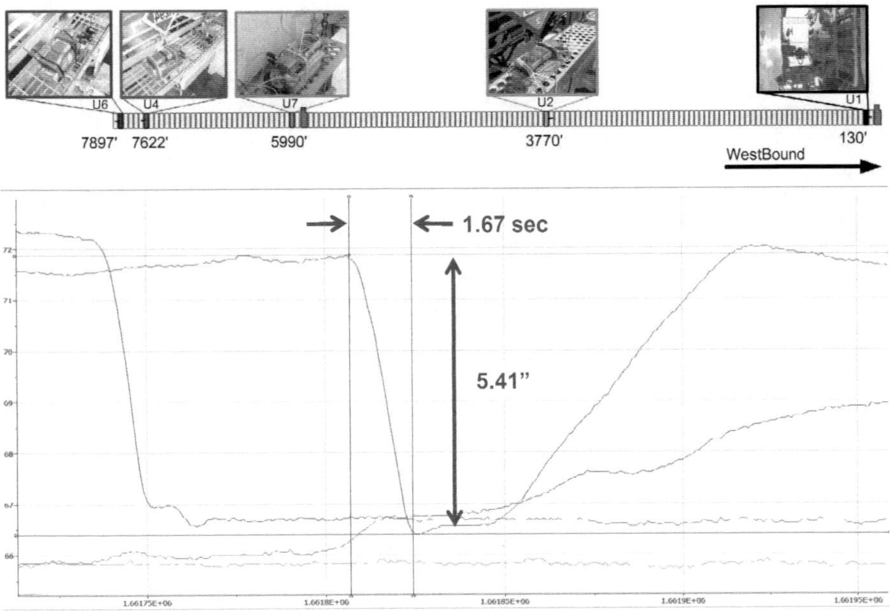

… half way through 330 Mile trip and 8.5 G-byte of data, can measure 1/100" at 1/100th

imagination at work 25 /

Excellent TO AIDP Control

(of available locos)

- Further utilize/analyze individual Detailed Train Make-up data to identify potential "Whip Hammers" and adjust timing and/or magnitudes of DP asynchronous commands.
- Consider 1X0X1 configuration
- Refine / hold Train Marshalling Rules

imagination at work 26 /

Agenda

Trip Optimizer control of Top Of Rail lube

TO and Top Of Rail Lubrication

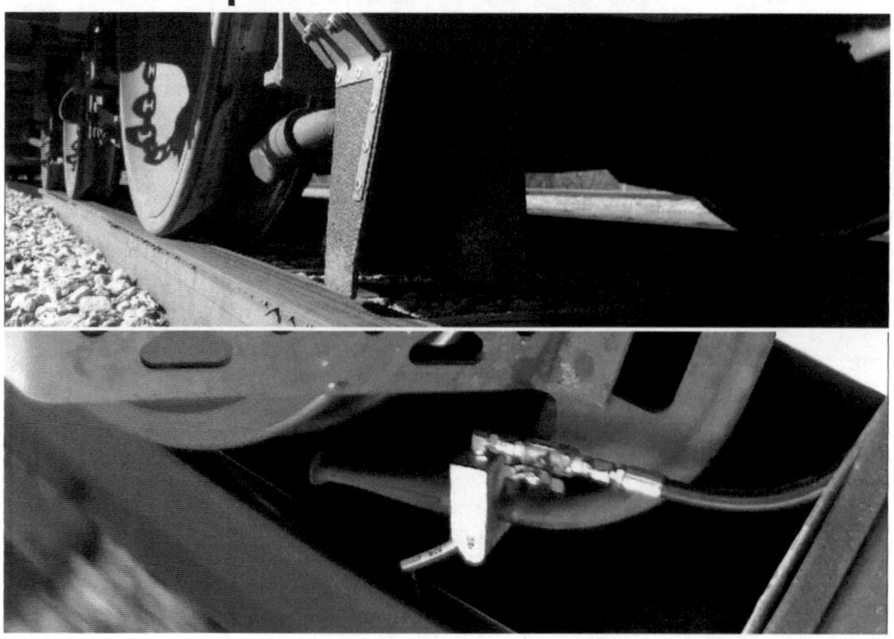

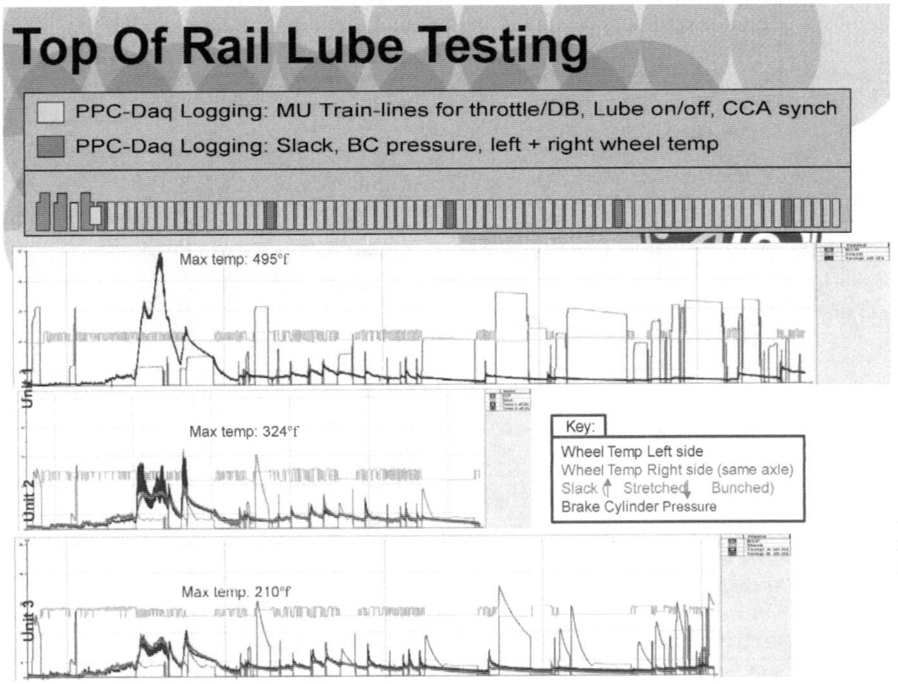

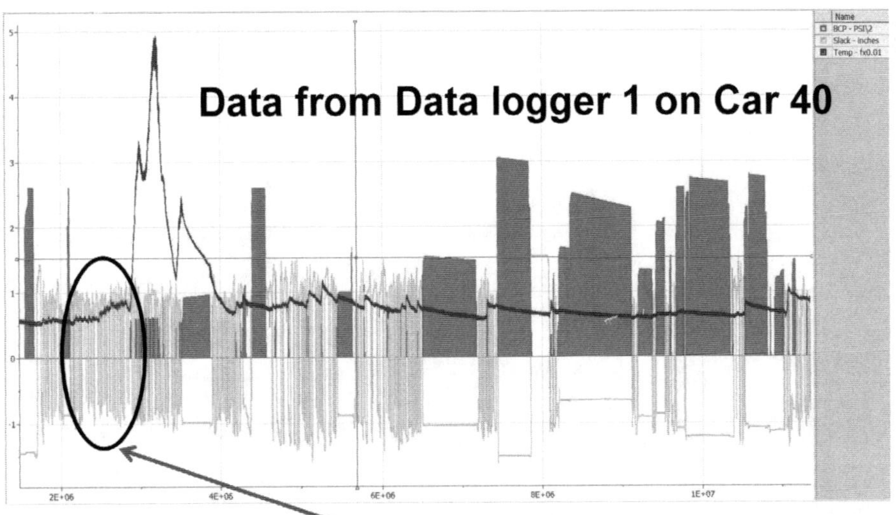

However, on long hill climb, data shows signs of significant difference in wheel heating with TOR on versus off…

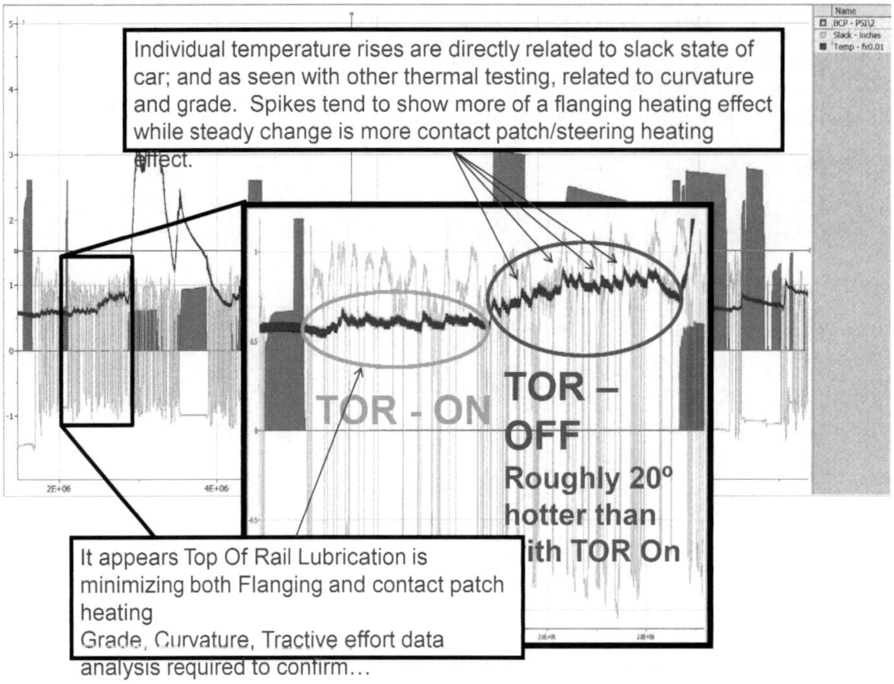
Individual temperature rises are directly related to slack state of car; and as seen with other thermal testing, related to curvature and grade. Spikes tend to show more of a flanging heating effect while steady change is more contact patch/steering heating effect.
Name
BCP - PSI\2
Slack - Inches
Temp - fx0.01
TOR - ON
TOR – OFF
Roughly 20° hotter than with TOR On
It appears Top Of Rail Lubrication is minimizing both Flanging and contact patch heating
Grade, Curvature, Tractive effort data analysis required to confirm…
2E+06
4E+06

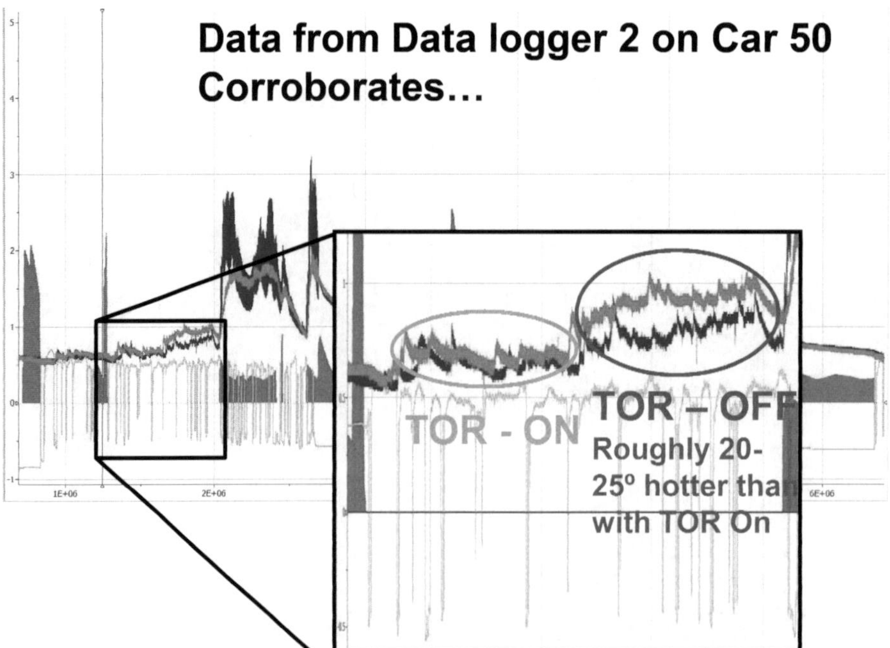
Data from Data logger 2 on Car 50 Corroborates…
TOR - ON
TOR – OFF
Roughly 20-25° hotter than with TOR On
1E+06
2E+06
6E+06

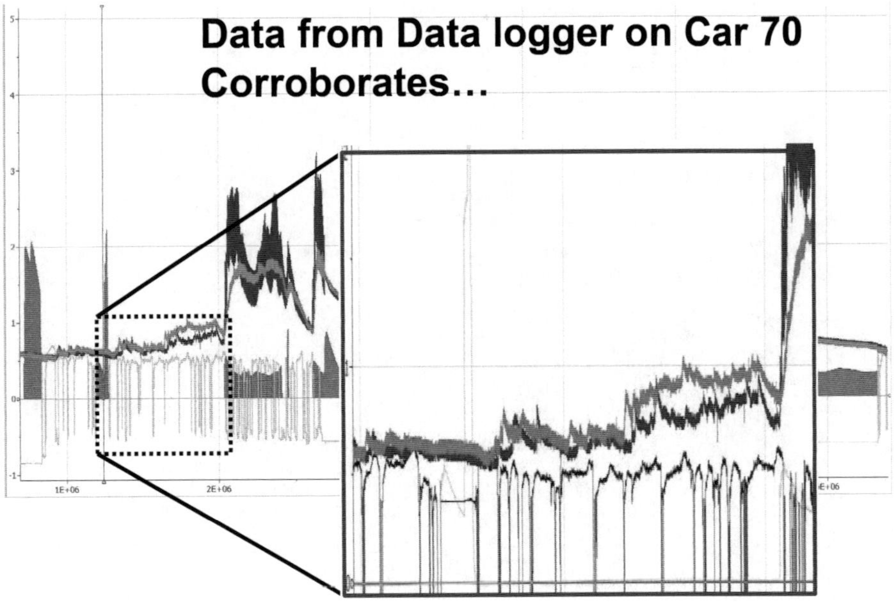
Data from Data logger on Car 70
Corroborates…
5
4
3
2
1
0
-1
1E+06
2E+06

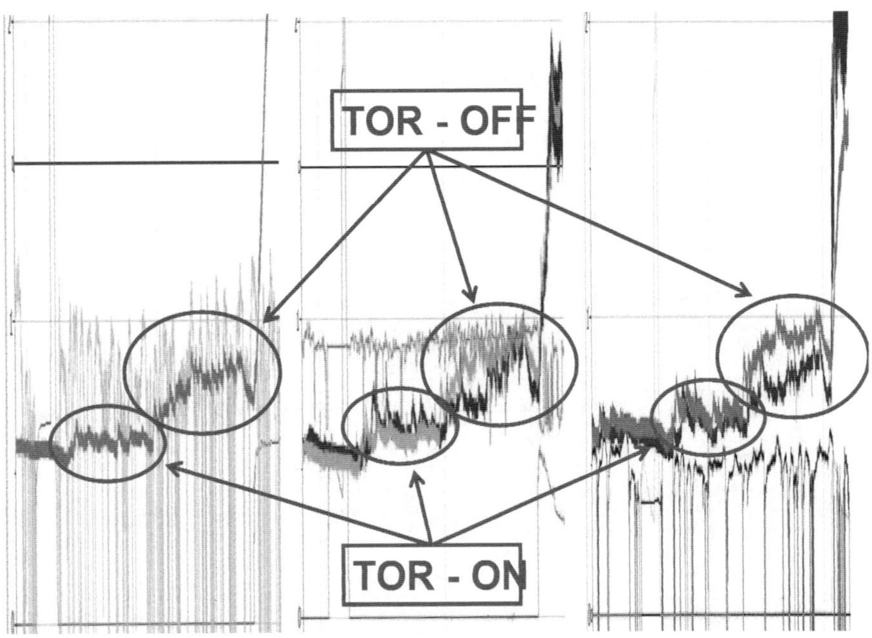
TOR - OFF
TOR - ON

Agenda

Trip Optimizer Advanced Algorithms for Optimization of Lateral Forces

W/R Interface vs Centripetal Force

Train Driving

Closest to natural steerage

Moving to flanging, disrupted steerage

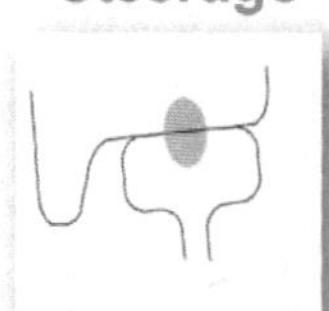

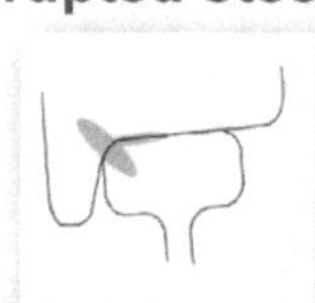

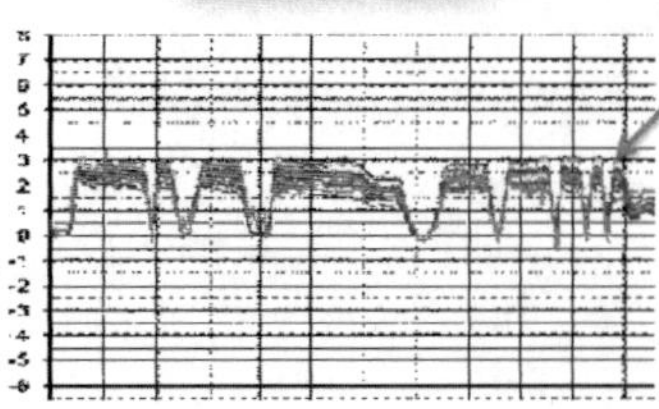

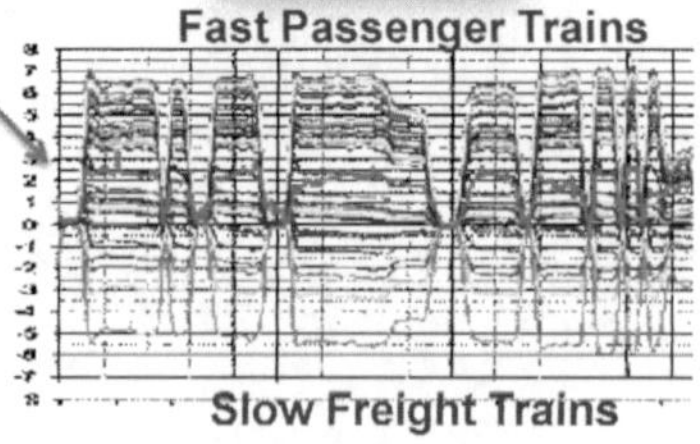

Controlling Train Speed to the Super-Elevation Speed

W/R Interface vs Centripetal Force

Train Driving

Minimal track and gauge impact

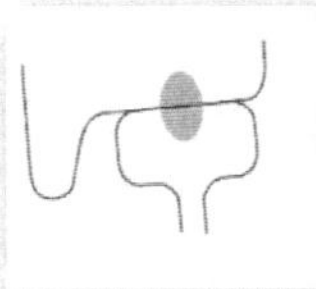

Gauge spreading and sleeper impact

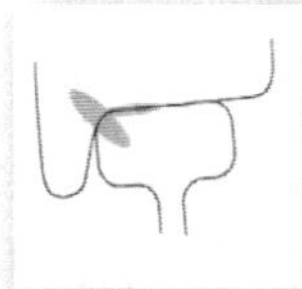

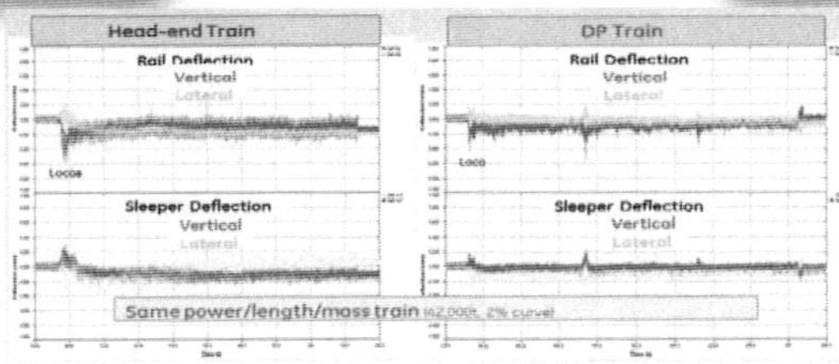

Controlling Train Speed to the Super-Elevation Speed

W/R Interface vs Longitudinal Force

Train Marshalling

Closest to natural steerage

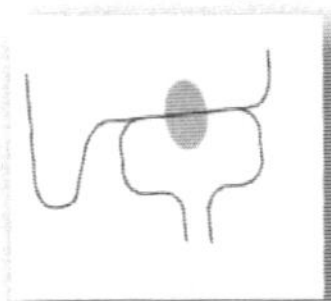

Moving to flanging, disrupted steerage

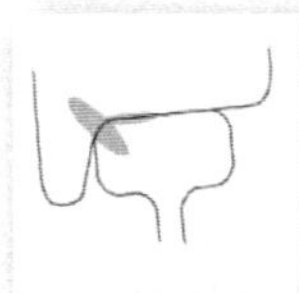

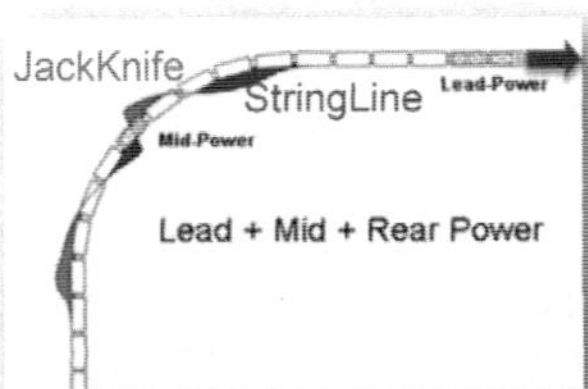

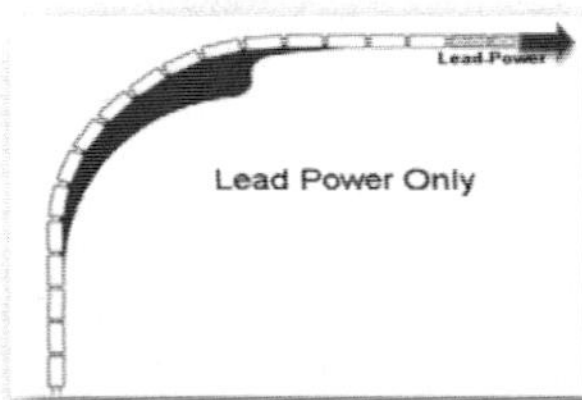

Distributing Power to Minimize Lateral Force

Kevin Koch, TTCI

Kevin is a Principal investigator for TTCI where he has been employed since 2001. Kevin holds a Bachelor of Science degree in Aerospace Engineering and is a licensed professional engineer. Kevin's work at TTCI has focused on all aspects of structural analysis from finite element analysis to fatigue analysis, accelerated fatigue testing, and the measurement of revenue service load environments.

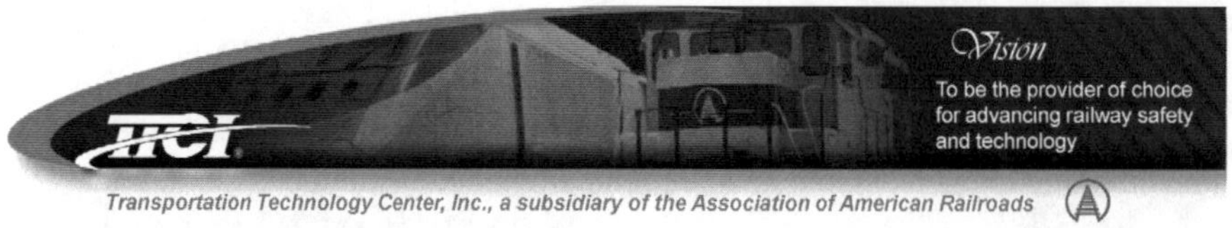

Transportation Technology Center, Inc., a subsidiary of the Association of American Railroads

TTCI Research
Coupler Force Environment and Coupler Durability

Kevin Koch
Principal Investigator

Coupling Component Structural Integrity

- **Problem Scope**
  - Between 2000–2009 over 80,000 knuckles and 56,000 couplers were removed because they were cracked or broken
  - Exact age unknown but most likely they were removed prior to acceptable service life of 20 to 30 years or 1.0- to 1.5-million miles
- **Objective of TTCI Coupler Component Research**
  - Determine nominal stress levels in knuckle & coupler castings
  - Determine current status of knuckle & coupler casting tolerances and influence of tolerances on component stress
  - Estimate fatigue life of critical coupler components when subjected to a representative service load environment
  - Estimate fatigue life of coupler components with typical surface defects when subjected to a representative load environment

TTCI Coupling Component Structural Integrity
Industry Survey of Broken Knuckles and Couplers

- **Survey Details**
  - Components included cracked and broken castings
  - Defects categorized by location on part
  - Most cracked knuckles found using magnetic particle technique found in AAR Specification M-220, developed by TTCI-NDT staff

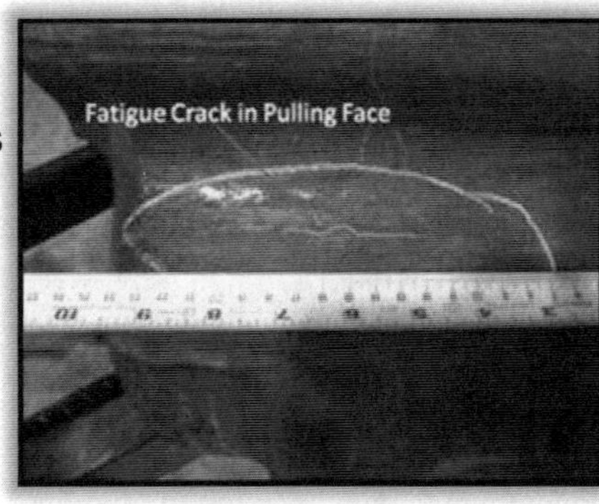

Fatigue crack in pulling face of knuckle found using the Magnetic Particle Inspection Method

TTCI Coupling Component Structural Integrity
Industry Survey of Broken Knuckles and Couplers

- **Survey Results – Coupler Body**
  - A total of 53 coupler bodies examined
  - Butt wear and deformation were the most common failure modes
  - Lug and pin hole cracks also frequently observed

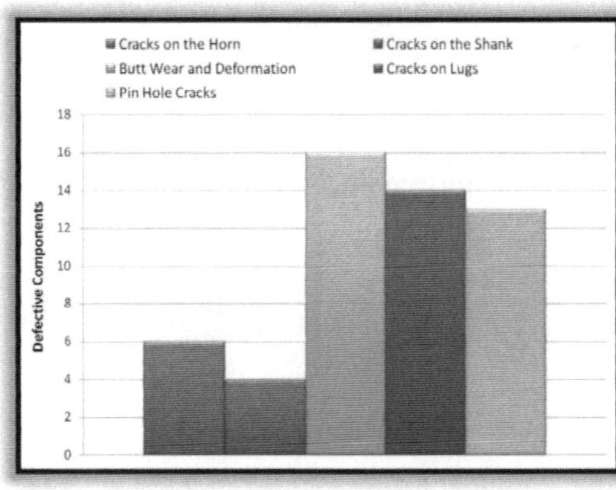

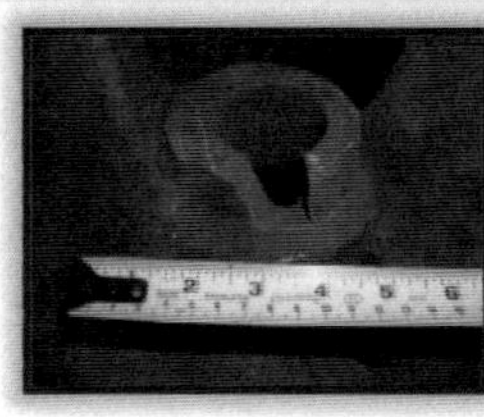

Coupling Component Structural Integrity

Industry Survey of Broken Knuckles and Couplers

- **Survey Results – Knuckle**
  - 68 knuckles examined
  - 64 of 68 knuckles contained fatigue cracks in pulling face
  - Cracking was also identified in the flag hole and lugs

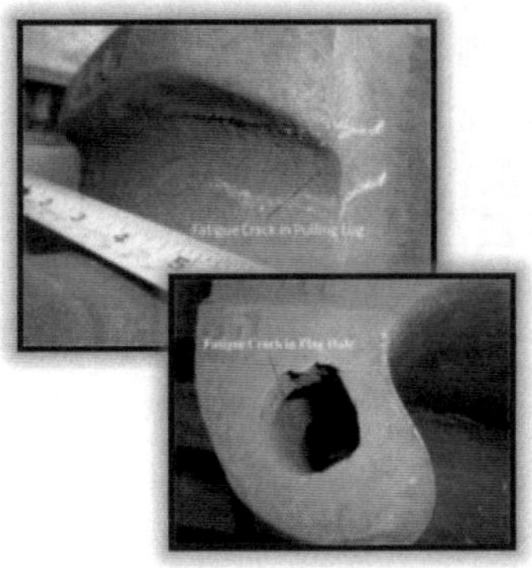

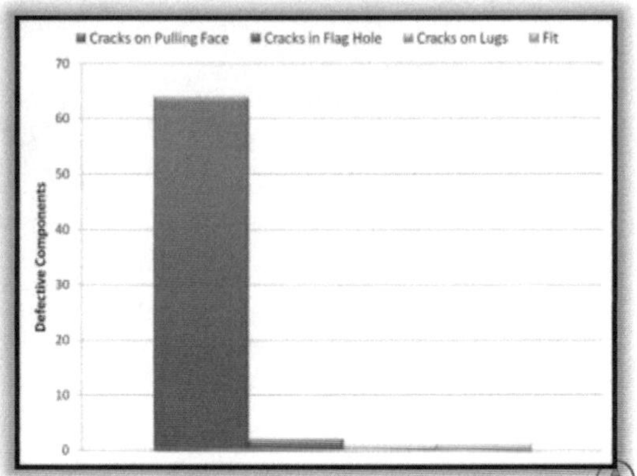

Coupling Component Structural Integrity

Knuckle and Coupler FEA

- **Knuckle & Coupler Stress Analysis**
  - Develop finite element models to analyze current designs using current load environment
- **Approach**
  - Use laser scanning technology to produce solid models for knuckle and coupler
  - Determine critical stress areas using Finite Element Analysis (FEA)
  - Compare stress results to strain gage results in critical stress areas

Coupling Component Structural Integrity
Knuckle and Coupler FEA

Modeled as Assembly

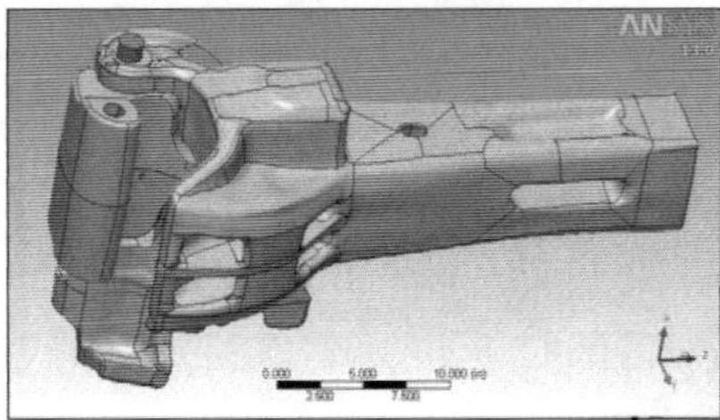

Contact Elements Between Knuckle & Coupler

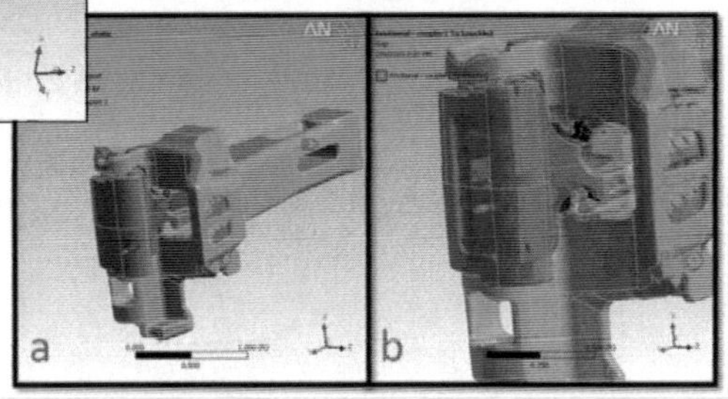

Coupling Component Structural Integrity
Knuckle and Coupler FEA

Element Mesh

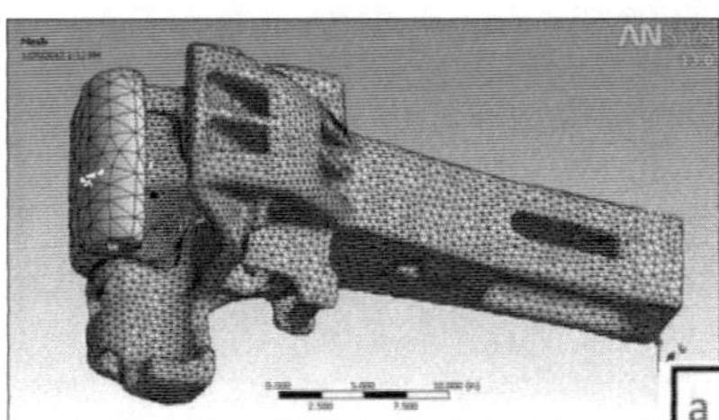

Finer Element Mesh at Pulling Lug

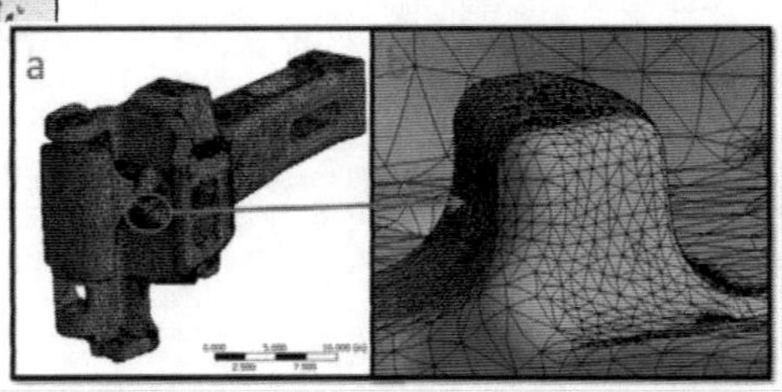

Coupling Component Structural Integrity
Knuckle and Coupler FEA

- Original model showed misalignment in pin hole (1/8 inch) of coupler body
- Models were modified to determine the influence of the misalignment on contact conditions of coupler lugs

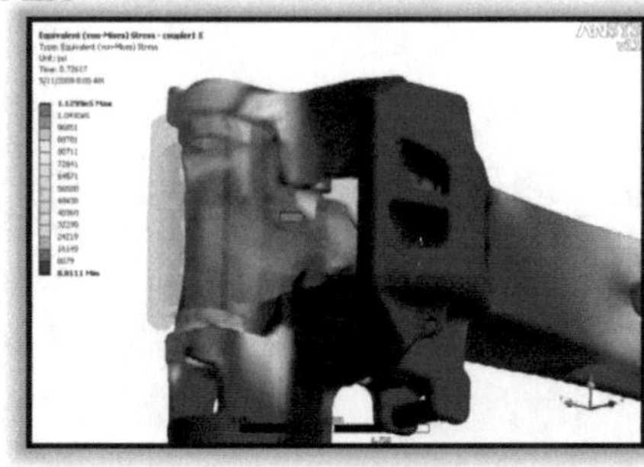

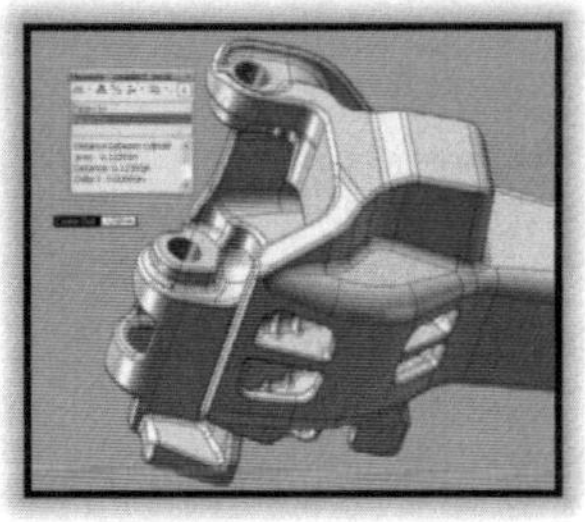

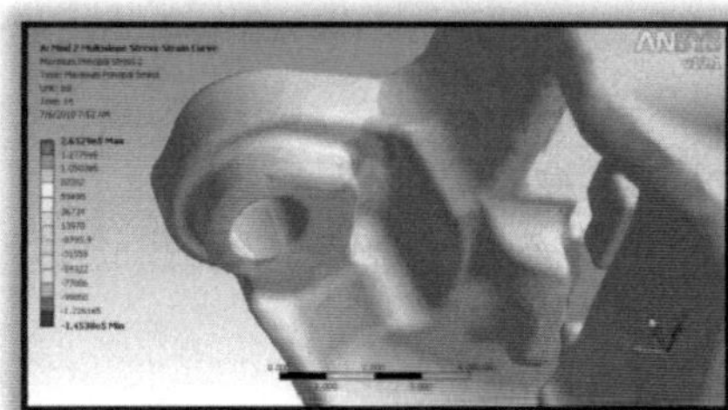

Coupling Component Structural Integrity
Knuckle and Coupler FEA

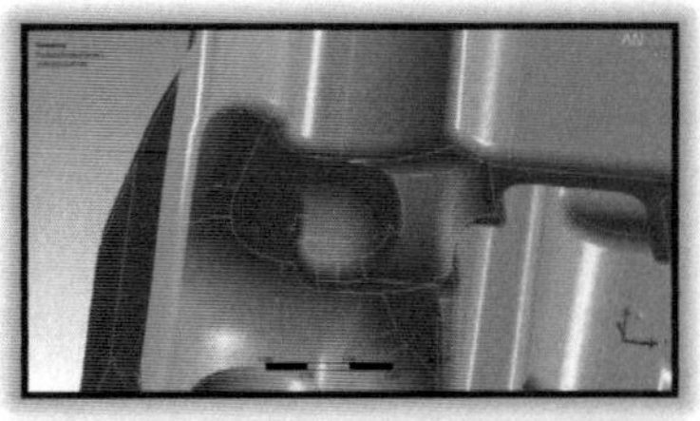

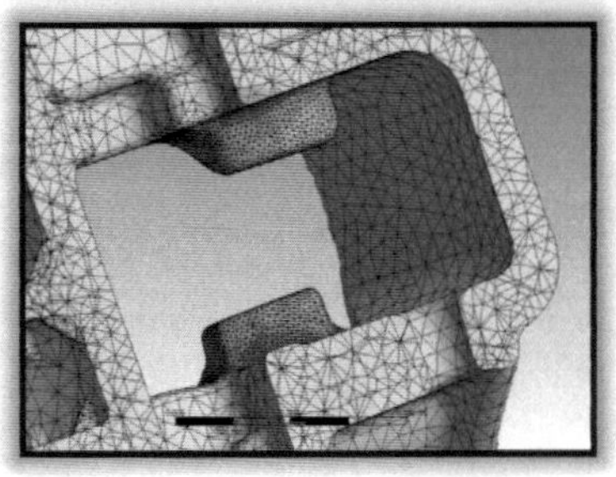

- **Correction to Model Dimensions**
  - Model modified to align pin holes
  - Significant draft load between knuckle and coupler shifted away from pin and into both restraint lugs
  - Simulation results more closely reflect those of strain gage test results with actual knuckle and coupler

Coupling Component Structural Integrity

Knuckle and Coupler FEA

- **FEA Results – Manufacturer B**
  - AAR M201 Gr. B – Yield 100,000 to 110,000 psi
  - At 240,000-pound draft load:
    - Top pin support – 113,500 psi
    - Bottom pulling lug – 97,000 psi
    - Knuckle face – 91,000 psi
  - **Top pulling lug not reacting significant load**

Stress vs. Draft Load

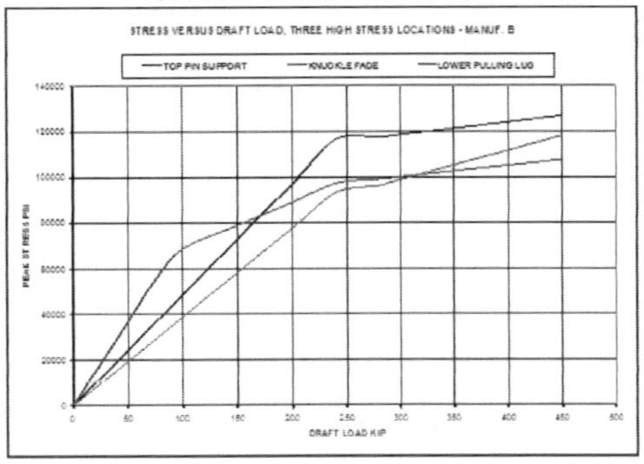

Coupling Component Structural Integrity

Knuckle and Coupler FEA

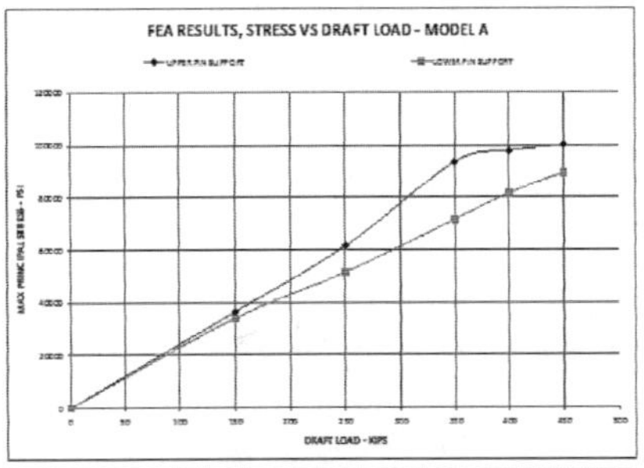

◄Pin Supports

- **FEA Results — Manufacturer A**
  - At 250,000-pound draft load:
    - Top pin support – 61,500 psi
    - Bottom pulling lug – 104,000 psi
    - Knuckle face – 103,000 psi
  - Top pulling lug reaching yield at only 125,000-pound draft load

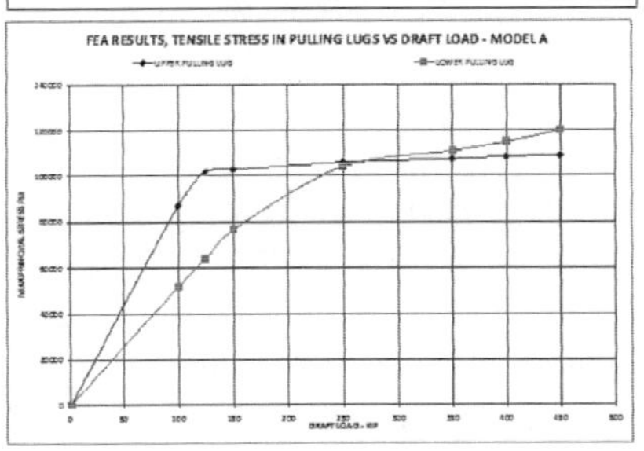

◄Pulling Lugs

Coupling Component Structural Integrity

Knuckle and Coupler Strain Gage Test

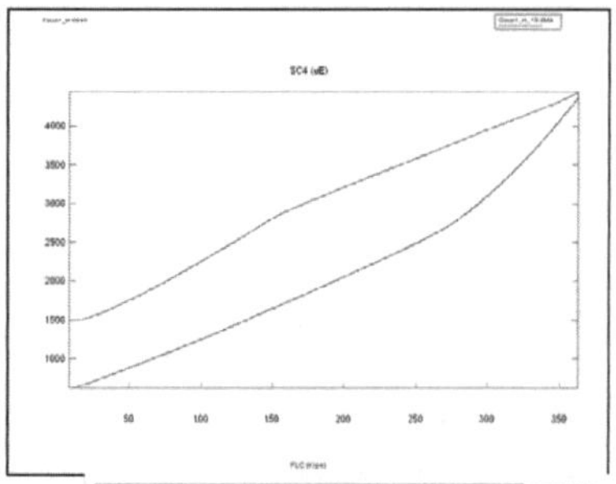

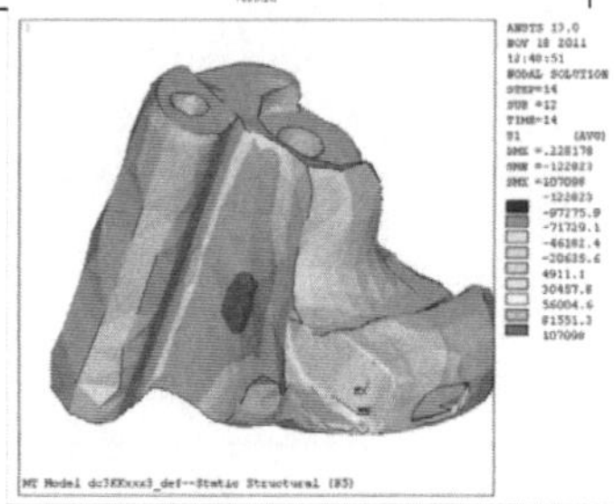

- **Test Results — Manufacturer A**
  - AAR M201 Gr. B – Yield 100,000 psi
  - At 250,000-pound draft load:
    - Top pin support – 63,500 psi
    - Bottom pulling lug – 80,000 psi
    - Knuckle face – 64,000 psi
  - **Bottom pulling lug yield at about 290,000-pound draft load**
  - **Yield detected in knuckle face at draft load of about 430,000 pounds**

Coupling Component Structural Integrity

FEA & Strain Gage Test Results Emphasized Need to Better Understand Scope of Variations of Key Coupler/Knuckle Assembly Dimensions

Improved Castings & Inspection Procedures

Current State of Tolerances

- Manufacturers use gages specified in M-211 or gages developed by the manufacturer under M-215
- Industry representatives and shop personnel report that new, as-manufactured components exhibit fitment issues
- Grinding on components is often required to ensure fitment between coupling system components

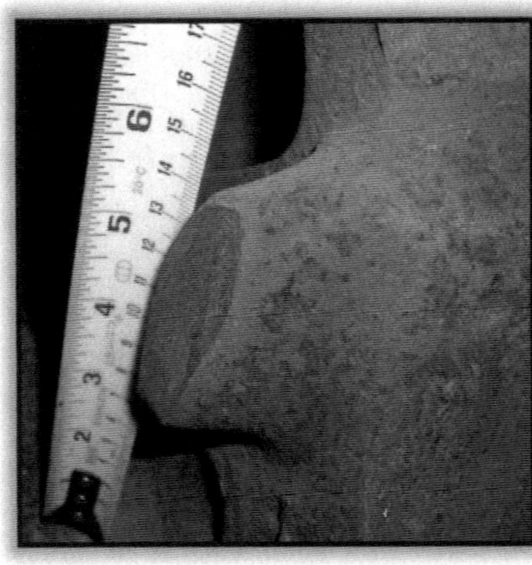

Improved Castings & Inspection Procedures

- **Since 2009, a total of 31 knuckles and 21 coupler bodies have been scanned**
  - Railroads supplied components from current stock
  - First set of data collected on E-type assemblies
- **Data indicates that coupler tolerances can be improved**

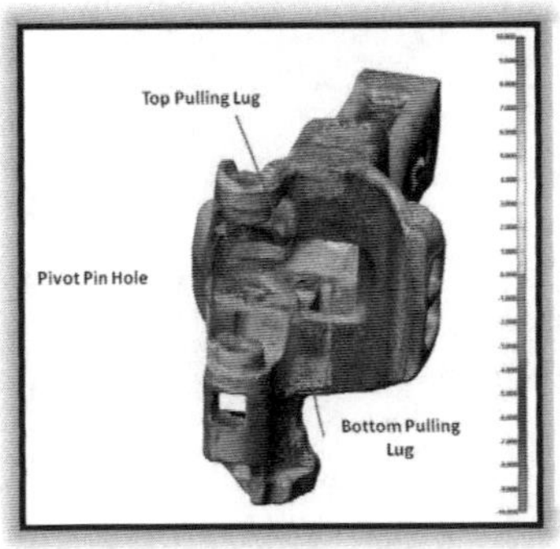

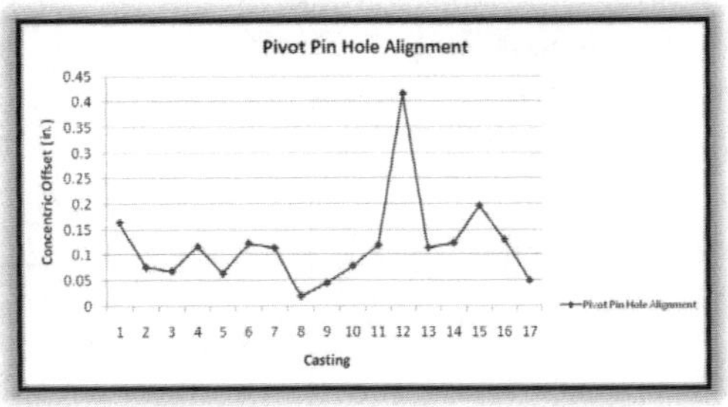

Standard Deviation by Manufacturer

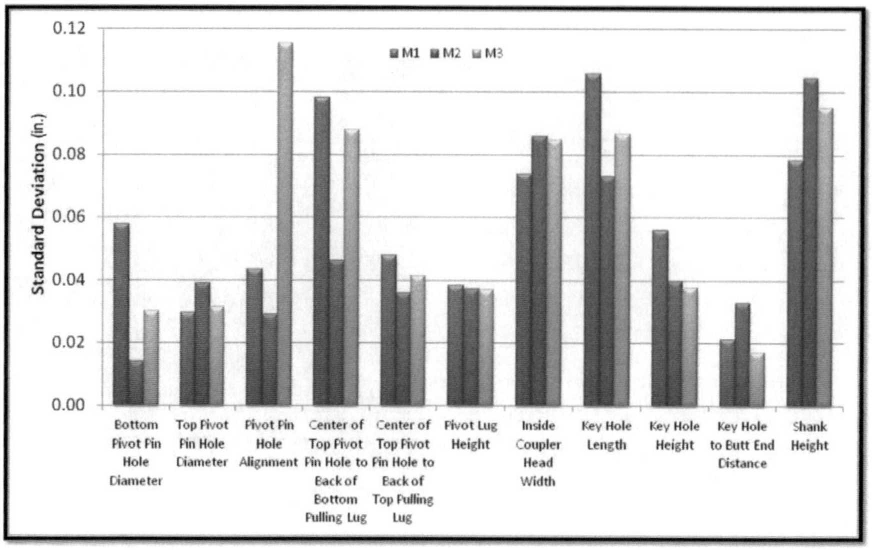

Critical Dimension — Pivot Pin Hole Alignment

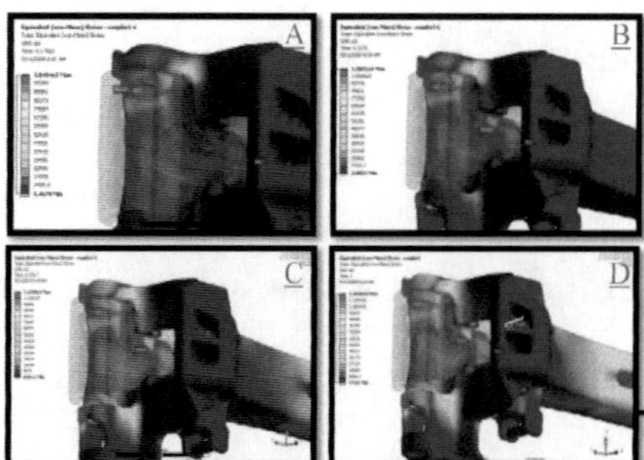

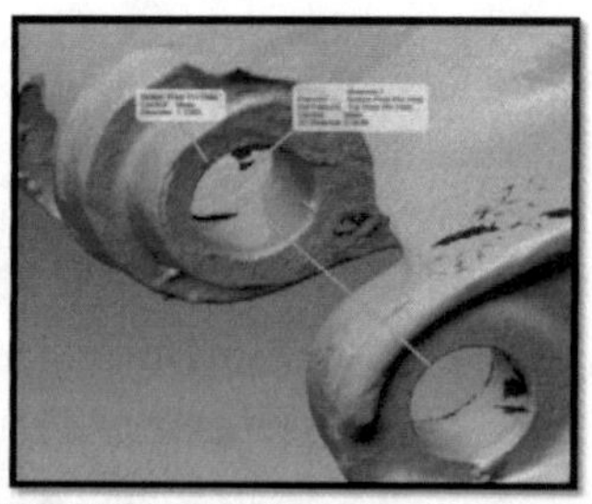

Pivot Pin Hole Alignment		
Average		0.105
Max		0.416
Min		0.019
Standard Deviation	M1	0.0435
	M2	0.0291
	M3	0.1155

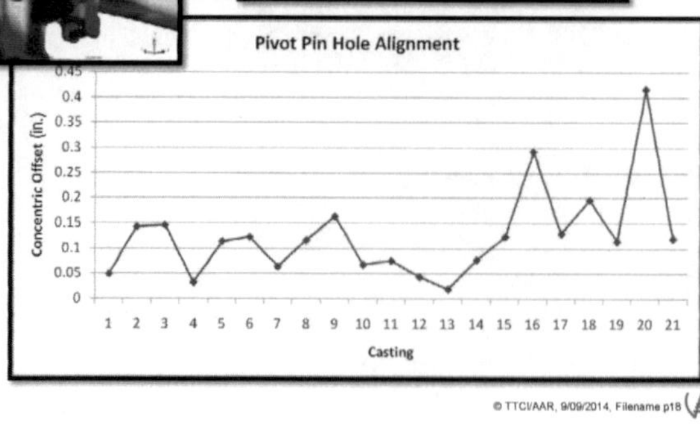

Critical Dimension: Pivot Pin Hole to Back of Pulling Lugs

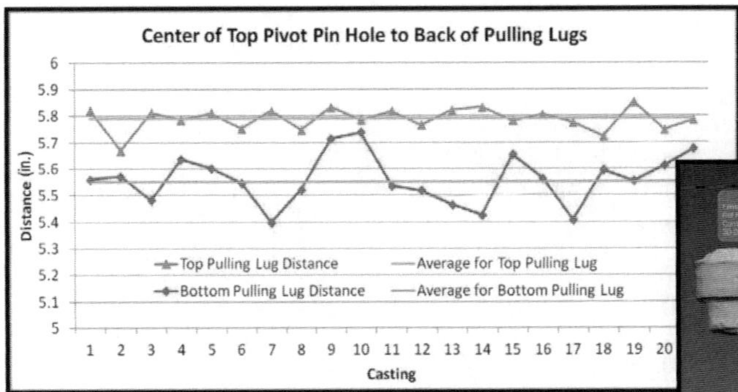

	Average	Maximum	Minimum	Standard Deviation
Center of Top Pivot Pin Hole to Back of Bottom Pulling Lug	5.5493	5.7385	5.3976	0.0939
Center of Top Pivot Pin Hole to Back of Top Pulling Lug	5.7908	5.8520	5.6707	0.0430

Coupler Geometry Review

- Coupler pulling lugs

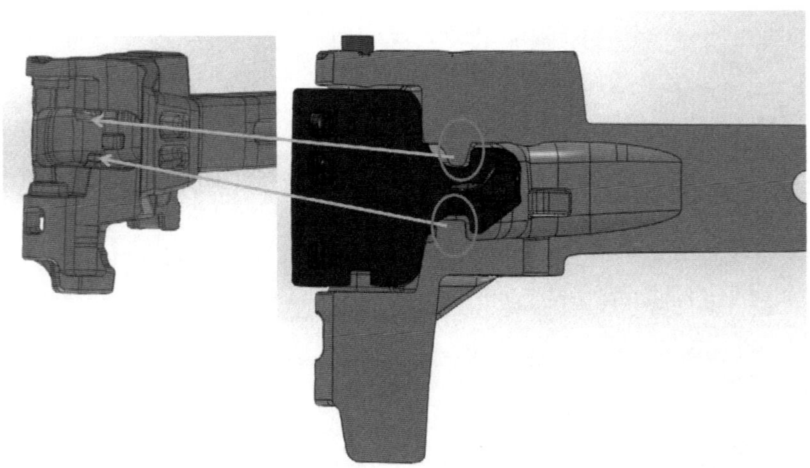

Coupler Geometry Review

- Knuckle lugs

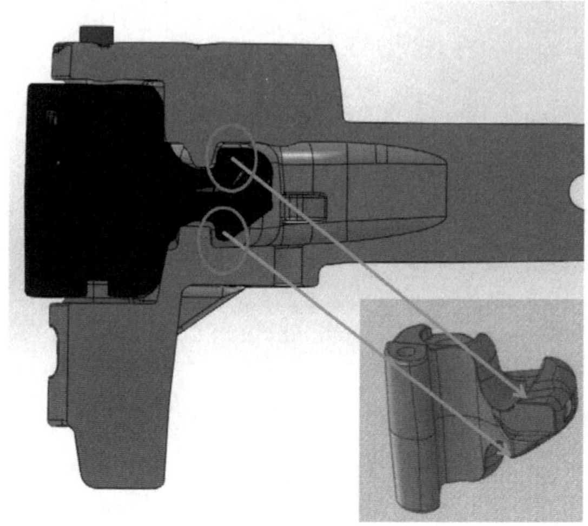

Finite Element Modeling

- Key measurements that can produce high principle stresses

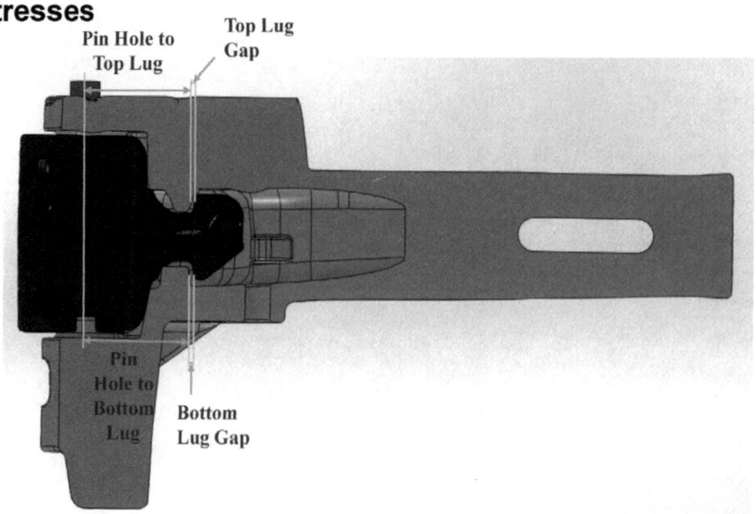

Coupler Geometry Review

- Other load transfer points

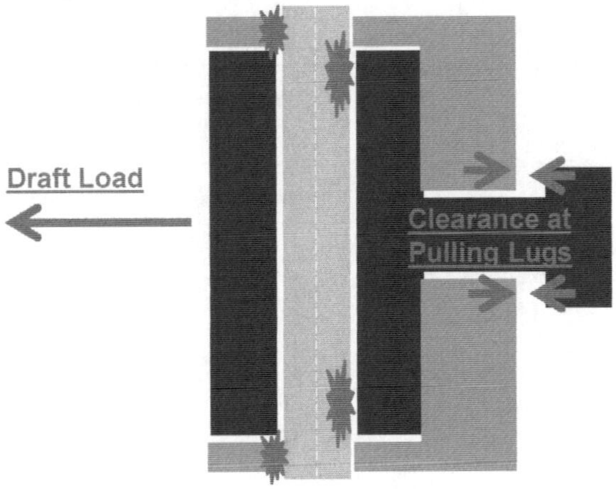

Case 1: Bottom Coupler Pulling Lug

- **Bottom coupler pulling lug moved to maximum variation location**
  - Results in tighter fit between knuckle/coupler lugs
- **Model inputs**
  - Includes average knuckle, lock, and pin
  - Only one dimension varied per model
  - Restraints applied to the pin and the coupler shank
  - 450,000-pound draft load applied at knuckle pulling face

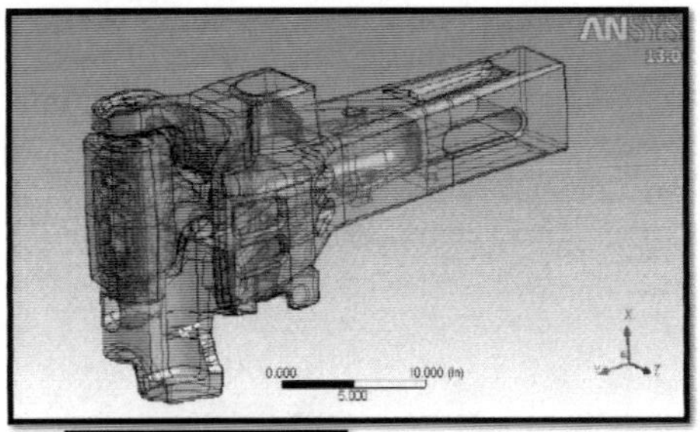

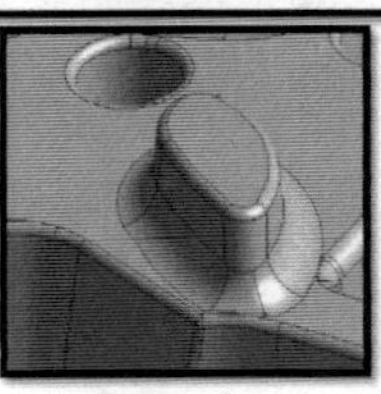

Bottom Lug Maximum Position

Bottom Lug Nominal Position

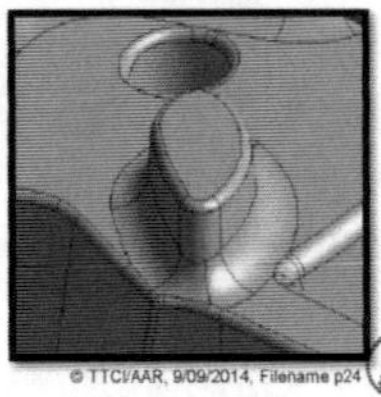

E-Type Knuckle Dimensional Analysis

- Effects of gap between pulling lug gaps on lug stress versus draft load

Knuckle/Coupler Combination	Knuckle To Coupler Gap Upper Pulling Lugs	Knuckle To Coupler Gap Lower Pulling Lugs
Manufacturer A	0.1162	0.2151
Manufacturer B	0.2113	0.2579
Manufacturer B Mod	0.3342	0.1706
Manufacturer B Mod 2	**0.1213**	**0.2547**

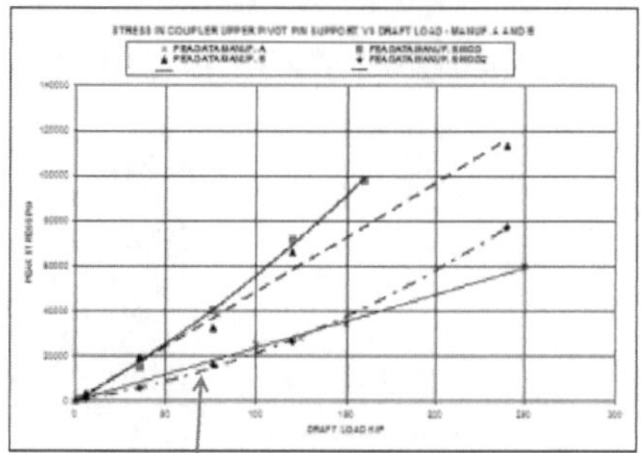

▲**Upper Pin Support**

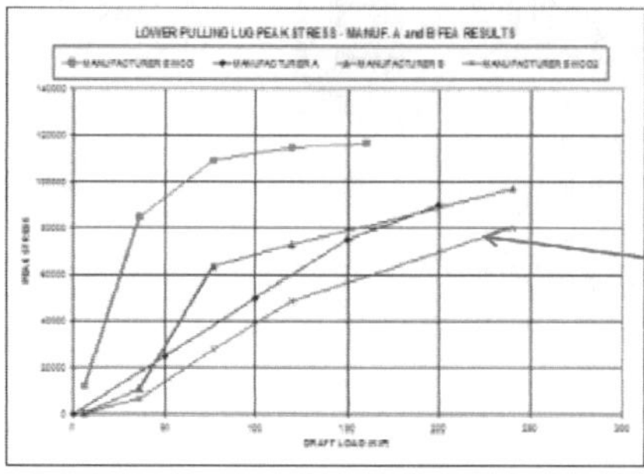

Manufacturer B Mod 2

◀**Lower Pulling Lugs**

Finite Element Modeling Results

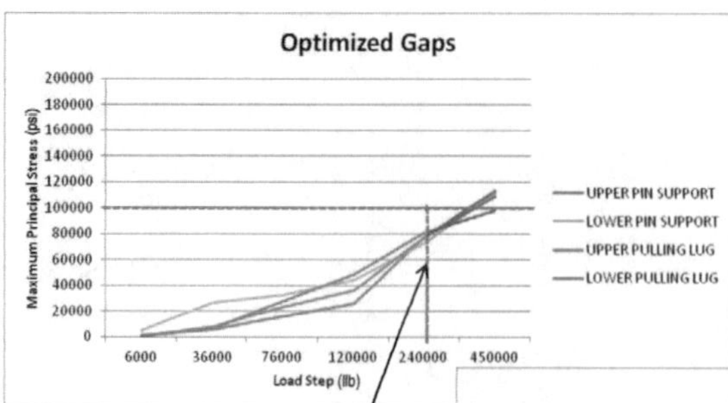

- Yield for current E-type material is 100,000 psi
- Optimized gaps model reduces maximum principal stresses significantly

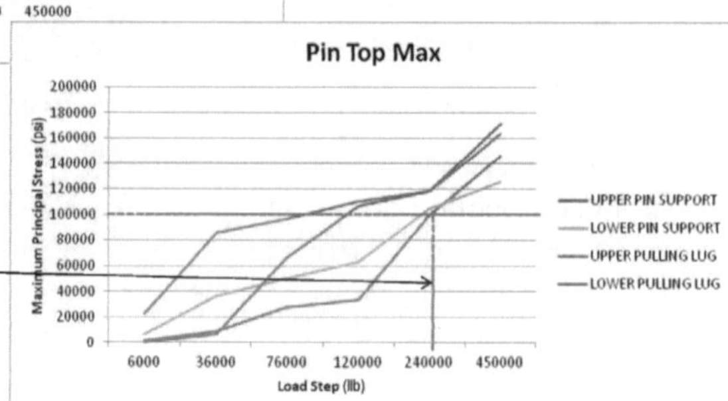

Typical high draft load seen in revenue service

E-Type Gages

- Critical dimensions from optimized model used in gages
- Estimated tolerances in critical areas are .012 inch per inch
- Two types of gages
  - Digital Gage
    - Requires laser scanner
    - Requires coupler to be cut in half
    - Used to take measurements at foundry
  - Physical Gage
    - Can be used in shops

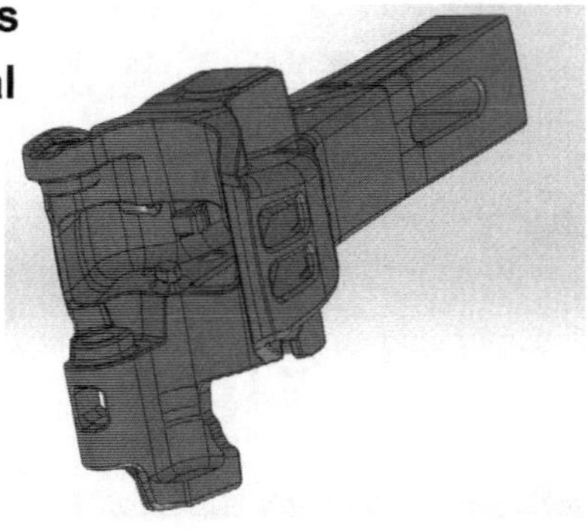

Coupler and Knuckle Fatigue

QUESTION: How do stress levels in knuckle and coupler relate to fatigue life of those components?

Improved Knuckle & Coupler Durability

Knuckle and Coupler FEA

- **Fatigue Analysis**
  - Relationships between coupler force and local stress established
  - Gouge and void type defects added to high stress areas in knuckle and coupler
  - Results from defect analysis used to produce fatigue notch factors for the fatigue life estimates

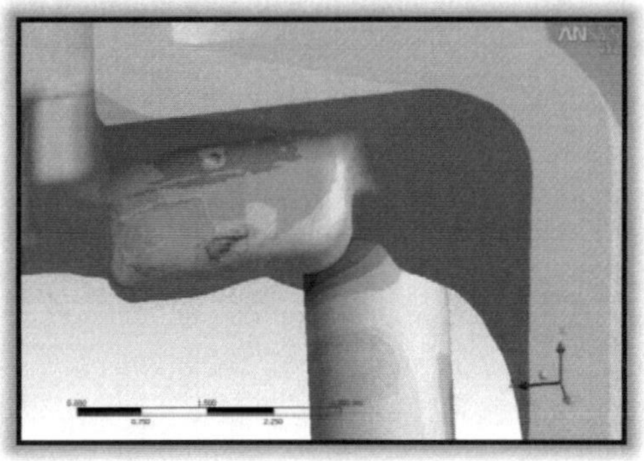

© TTCI/AAR, 9/09/2014, Filename p29

Knuckle and Coupler Fatigue Life Estimation

Fatigue Life Estimates for Loaded Hopper Car – 68% Survival (Target Life of 1.5E6 Loaded Miles)			
Component	**Location**	**Defect Type**	**Estimated Life Until Crack Initiation (miles)**
Coupler	Upper Pin Support	**None**	6.179E6 to 6.968E6
Coupler	Upper Pin Support	0.1875 Gouge	2.310E6 to 2.476E6
Coupler	Lower Lug	**None**	**1.861E5 to 2.225E5**
Coupler	Lower Lug	0.1875 Gouge	**1.147E5 to 1.299E5**
Coupler	Lower Lug	0.0625 Void	**1.437E5 to 1.654E5**
Knuckle	Top surface	**None**	**1.404E6 to 1.647E5**
Knuckle	Pulling Face	**None**	1.604E6 to 1.896E6
Knuckle	Pulling Face	0.1875 Gouge	**2.895E5 to 3.160E5**

© TTCI/AAR, 9/09/2014, Filename p30

Coupler Structural Durability

- **Report Summarizing Coupler Research**
  - AAR Research Report: "Comprehensive Evaluation of the Structural Durability of Railroad Coupler Assemblies" R-1002, August 2012, K. Koch & K. Gonzales

Coupler Structural Durability

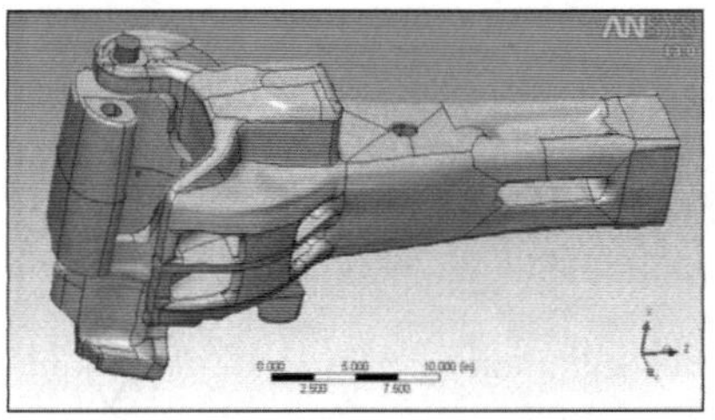

Questions?

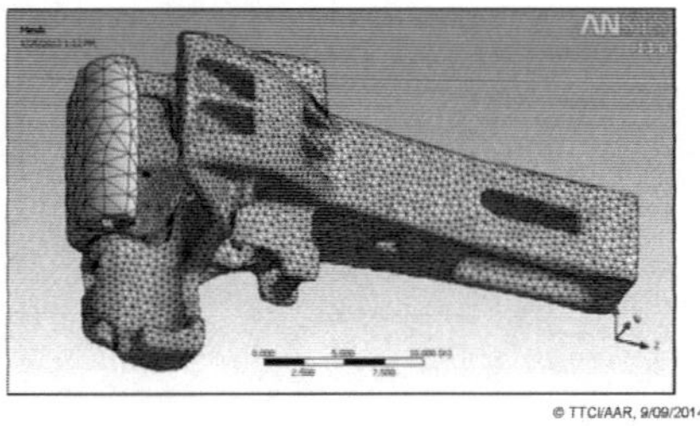

Coupler/Knuckle Structural Durability

Coupler Load Environment For
Cushion Unit Equipped Freight Cars

Additional Dynamic Load Environment Testing, Cushion Unit Cars — 2012 to 2014

Test Car 1: 286,000-pound GRL Box

Payload – Water Filled Barrels & Steel Shapes

End of Car Cushion Unit

Additional Dynamic Load Environment Testing, Cushion Unit Cars — 2012 to 2014

TEST DISTANCE – TOTAL 21,220 MILES		
Railroad	**Routes**	**Distance (miles)**
Western	General Service - Colorado to Alabama, and California to Chicago	**4,580**
Western	General Service - Chicago to Washington, Washington to Colorado	**2,785**
Western	Unit Coil Train – Alabama to California	**1,870**
Northern	General Manifest Service in Ontario and Quebec	**11,985**

Additional Dynamic Load Environment Testing, Cushion Unit Cars — 2012 to 2014

CUSHION UNIT USAGE DURING REVENUE SERVICE TEST		
Cushion Unit	**Routes**	**Distance (miles)**
Standard Used	**Western and Northern**	**12,241**
Standard New	**Northern**	**1,926**
Active Draft	**Western and Northern**	**6,990**

Additional Dynamic Load Environment Testing, Cushion Unit Cars — 2012 to 2014

Standard & Active Draft Cushion Units

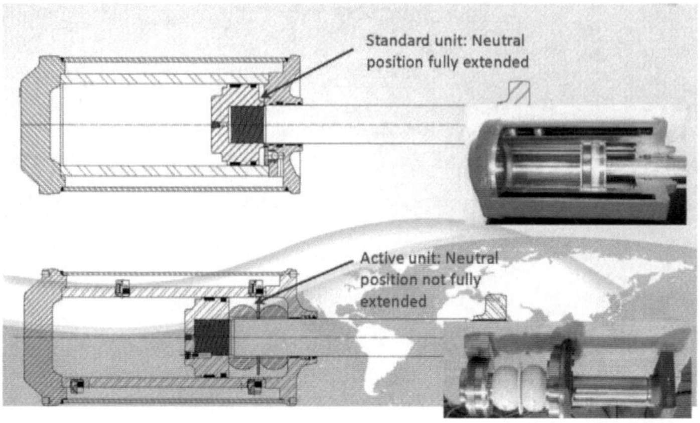

© TTCI/AAR, 9/09/2014, Filename p37

Additional Dynamic Load Environment Testing, Cushion Unit Cars — 2012 to 2014

- Data Processing & Analysis
  - Time domain files reviewed when high load event indicated.
  - Coupler force time domain data rainflow cycle counted
  - Method to compare coupler load severity — knuckle fatigue damage due to draft load
  - Closer investigation was conducted for events indicating especially high draft load and coupler displacement activity: locomotive positions, consist details and length, train speed

Knuckle Model

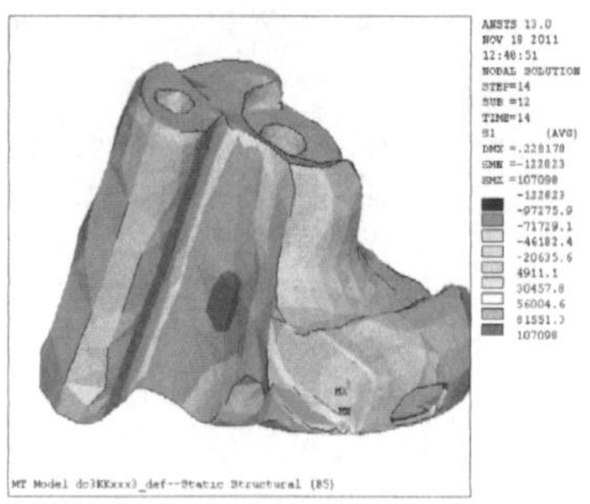

© TTCI/AAR, 9/09/2014, Filename p38

Additional Dynamic Load Environment Testing, Cushion Unit Cars — 2012 to 2014

CALCULATED KNUCKLE DAMAGE – EXISTING AAR SPECTRA Relative Knuckle Fatigue – Damage per Mile	
Load Spectrum	**Relative Damage**
AAR Hopper Car "Severe"	1.00
AAR M-216 Test Spectrum	1.22

Relative damage number of 1.0 equivalent to a life until crack initiation on the knuckle face of about 6.69 to 9.14 million miles depending on material quality. Presence of surface defects could reduce life to about 1.6 to 1.7 million miles

Additional Dynamic Load Environment Testing, Cushion Unit Cars — 2012 to 2014

- Coupler Force Event – Train Action, Northern Route

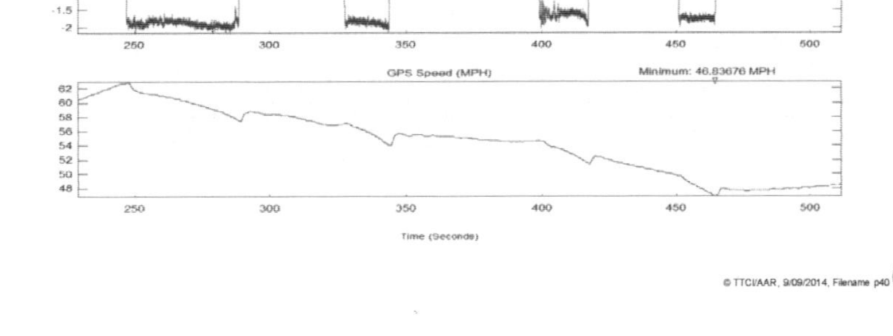

Coupler Force Environment — Cushion Unit Freight Car

Knuckle Damage per Mile – Standard 15-inch Cushion Units – Extended Distance

Cushioning Units	Type of Service	Test Miles	Relative Damage
Standard Used	General Freight & Unit Trains	12,241	0.44
Standard Used	Unit Coil Train	1,870	1.33
Standard Used	General Freight, Northern	5,854	0.86
Standard New	General Freight, Northern	1,926	3.97
Standard New & Used	General Freight & Unit Train, Western	14,790	1.00
Standard New & Used	General Freight, Northern	7,780	1.56

Knuckle Damage per Mile – Active Draft Cushion Units – Extended Distance

Type of Service	Test Miles	Relative Damage
General Freight Northern	4,205	0.54
General Freight Western	2,785	0.84

Coupler Force Environment — Cushion Unit Freight Car

Knuckle Damage per Mile – Single Trips, Northern RR Track – Test Car in Rear Third of Train Among other Loaded Cushion Unit Cars

Cushioning Units	Total Cars in Train	Test Car Position	Data Miles	Relative Damage
Standard Used	131	128	350	7.03
Standard Used	75 to 87	60 & 54	189	8.92
Standard Used	110*	85	177	10.30
Standard New	113	113	350	3.95
Standard New	139	118	139	5.59
Active Draft	129	127	353	5.36

*Test car 6 cars aft of 2 locomotives inserted near the center of the train. Different route.

Coupler Force Environment — Cushion Unit Freight Car

Relationship of Test Car to Other Cushion Unit Equipped Cars

Cushioning Units	Total Cars in Train	Size of Block of EOC Cars	Position of Test Car in EOC Block	Relative Damage
Standard Used	114	13	5 of 13	12.34
Standard Used	110*	15	6 of 15	10.30
Standard Used	75 to 87	13	7 of 13	8.92
Standard Used	131	20	17 of 20	7.03
Standard New	118	45	44 of 45	5.59
Active Draft	129	70	68 of 70	5.36
Standard New	113	38	38 of 38	3.95

*Test car 6 cars aft of 2 locomotives inserted near the center of the train. Different route.

Coupler Force Environment — Cushion Unit Freight Car

Knuckle Damage per Mile – Single Trips, Northern RR Track – Test Car in Middle to Rear Third of Train among Empty Cushion Unit Cars or Cars with Standard Draft Gear

Cushioning Units	Total Cars in Train	Test Car Position	Data Miles	Relative Damage
Standard Used	121	86	350	0.48
Standard New	65	41	350	0.59
Standard New	102	58	360	2.26
Standard New	118	48	350	2.51
Active Draft	121	59	350	0.79
Active Draft	156	83	350	2.65
Active Draft	135	78	350	1.91

Coupler Force Environment — Cushion Unit Freight Car

- **Examination of the data revealed the following key observations:**
  - Maximum train action related draft force was 460,000 pounds.
    - Draft forces as high as 225,000–275,000 pounds were also relatively common throughout the duration of the test.
  - Maximum train action related buff force was 320,400 pounds.
    - However, peak buff forces greater than ~225,000 pounds were relatively rare.
  - During periods of significant train action, cushion unit displacement was recorded over a range of ~0.8-inch extension to 2.0 inches compression.

Coupler Force Environment — Cushion Unit Freight Car

- **Key Observations/Conclusions**
  - Coupler force environment recorded over long distances and a wide variety of operating conditions:
    - Little significant difference in knuckle damage per mile, cushion units vs. standard draft gear.
  - Coupler force environment recorded with test car in middle to rear third of train and coupled to cars with standard draft gear or empty cushion unit cars:
    - Fatigue damage could increase depending on train handling but not enough to lead to unacceptable life.
  - Coupler force environment recorded with test car in rear third of train coupler among loaded cushion unit cars:
    - Fatigue damage could increase significantly over certain terrain, possibly leading to early failures with extended exposure.

Coupler Force Environment — Cushion Unit Freight Car

- **Key Observations/Conclusions – Isolation of Variables Not Sufficient & Results Inconclusive**
  - Active Draft Cushion Units
    - No clear evidence thus far that use of active draft cushion units provided a significant decrease in load environment when compared with standard cushion units.
  - Train Length
    - Train length is not identified as a major variable in draft coupler force environment severity.
  - Distributed Power
    - Limited comparisons did not identify any clear advantages of distributed power in reduction of draft coupler force environment severity.

Coupler Force Environment — Cushion Unit Freight Car

- **Reports Summarizing Boxcar Test Results**
  - *Technology Digest*: TD-13-027, released November 2013
  - Research Report: R-1006, released June 2014

Coupler Force Environment — Cushion Unit Freight Car

Second Test Car for Revenue Service Testing

- Bi-level autorack car with 10-inch standard cushion units
- Inserted primarily in unit train service with similar loaded autorack cars
- Primarily western U.S. routes — testing began Oct. 2013
- Has accumulated about 22,000 miles of revenue service data
- Almost 12,500 miles of data processed

Coupler Force Environment — Cushion Unit Freight Car

CALCULATED KNUCKLE DAMAGE – EXISTING AAR SPECTRA Relative Knuckle Fatigue – Damage per Mile	
Load Spectrum	**Relative Damage**
AAR Hopper Car "Severe"	1.00
AAR M-216 Test Spectrum	1.22

Knuckle Damage per Mile – Autorack Car & Boxcar Extended Distance

Car Type	Type of Service	Test Miles	Relative Damage
Boxcar	General Freight & Unit Train	12,607	0.50
Autorack Car	General Freight & Unit Train	12,428	1.71

Knuckle Damage per Mile – Autorack Car & Boxcar Severe Service

Car Type	Test Miles	Relative Damage
Boxcar	350	7.03
Autorack Car	396	7.39

Coupler Force Environment — Autorack Car

Typical Damaging Cycles – Coupler Force, Displacement & Car Speed

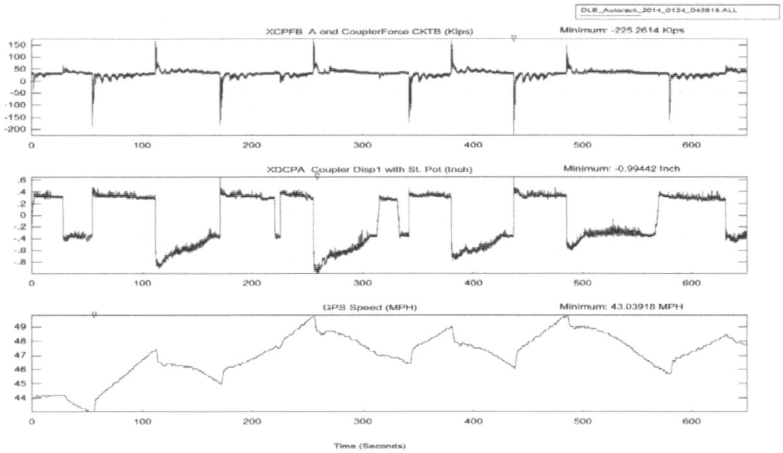

Coupler Force Environment — Cushion Unit Freight Car

Peak Buff Coupler Loads, Autorack Cars – Yard Impacts

Date	Time	Speed @ Impact (mph)	Peak Buff (lb)
Jan. 24, 2014	4:29 pm	0	846,400
Jan. 24, 2014	4:29 pm	0	705,300
Jan. 24, 2014	4:29 pm	0	641,900
Jan. 26, 2014	1:39 am	0	833,600
Jan. 27, 2014	12:33 pm	3	773,400
Jan. 28, 2014	2:02 pm	0	1,286,100
Jan. 28, 2014	2:12 pm	0	1,375,800
Jan. 29, 2014	7:12 am	0	750,500

Coupler Force Environment — Cushion Unit Freight Car

- **Reports Summarizing Autorack Test Results**
  - *Technology Digest*: forthcoming, October 2014
  - Research Report: forthcoming, December 2014

Coupler Durability & Force Environment

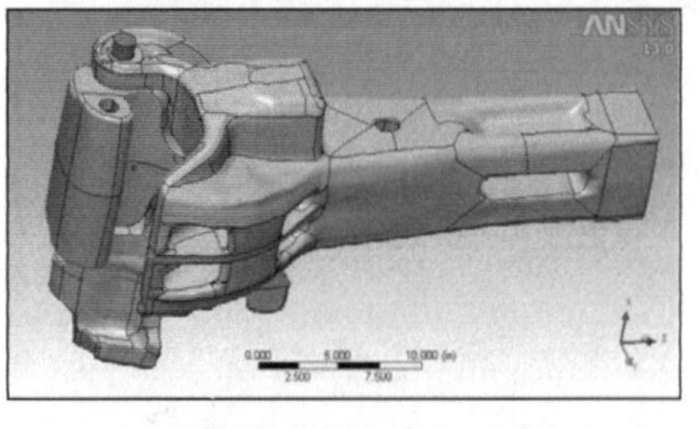

Questions ?

Dave Peltz, GE

Dave is the Chief of Global Technology for GE Transportation's Intelligent Control Solutions Division in Melbourne, Florida.

Dave's technical career began as an engineering consultant designing industrial assembly lines and automated control systems while in college.

With an Electrical Engineering degree, Dave joined the Air Force as a Systems Engineer for the MILSTAR satellite system, the military's nuclear war communications system. He invented network protocols and control algorithms for high efficiency nuclear missile re-targeting. He also designed the satellite comm distribution system for the Minute-Man and Peace-Keeper nuclear missile fields.

Dave was promoted to Program Manager of the Dual Modem upgrade program overseeing the $640 million Air Force Terminal Program source selection. Captain Peltz then went to Cape Canaveral to integrate, test, and launch satellites on the Space Shuttle and Titan rockets.

After the Air Force, Dave worked for Lockheed-Martin as Program Manager and Chief Communications System Engineer for the Eastern Launch Range, and designed the satellite comm system used for tracking or destruction of space craft launched from Florida.

Dave has been with GE Transportation for 18 years and has invented and advanced technology for products such as LOCOTROL Distributed Power, Remote Control Locomotives, the Trip Optimizer energy management system, ECP Brakes, and PTC. The in-train test equipment he has developed has significantly advanced the understanding of air brake performance and train dynamics. His brake system and pneumatics expertise and insight was instrumental in changing the long held 60 CFM departure rule to 90 CFM when using DP in Canada. Dave holds 31 U.S. and international patents in freight rail and communication

Longitudinal vs Lateral Train Force and Managing Wheel/Rail Interaction Through Train Operations

David Peltz
Chief, Global Technology
GE Transportation
Intelligent Control Systems

...Or "Can we all get along?"

Maintenance/Infrastructure
vs
Operations/Train Marshalling

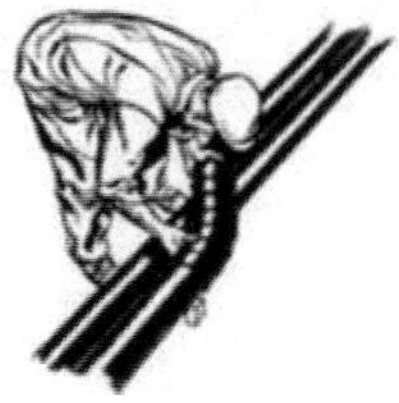

- **Often Conflicting Objectives (Need Champion of Both)**
- **Evaluate the Cost/Benefits (How? ...Data)**
- **Control the Operations/Marshalling**
(Maintenance Programs fairly well established)

W/R Interface vs Longitudinal Force

Train Marshalling

Closest to natural steerage

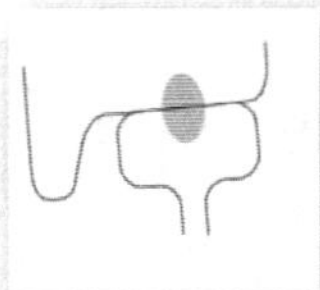

Moving to flanging, disrupted steerage

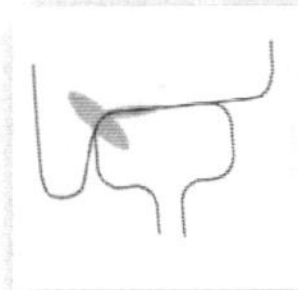

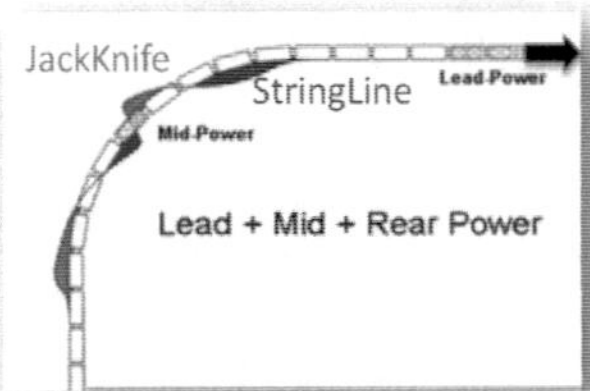

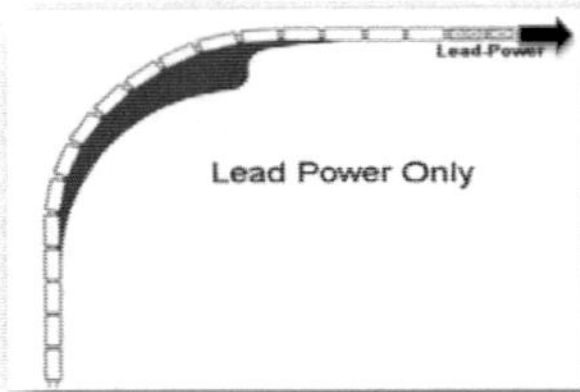

Distributing Power to Minimize Lateral Force

W/R Interface vs Slack Action

Both Driving and Marshalling

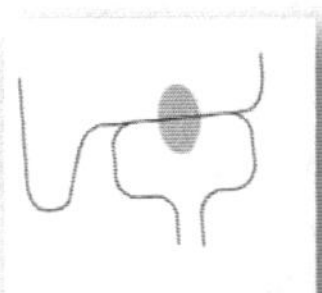

Rapid Transitions between flanging and disrupted steerage

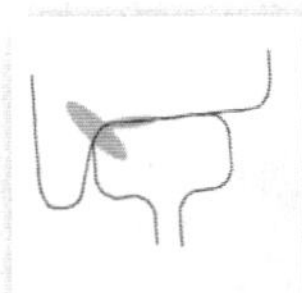

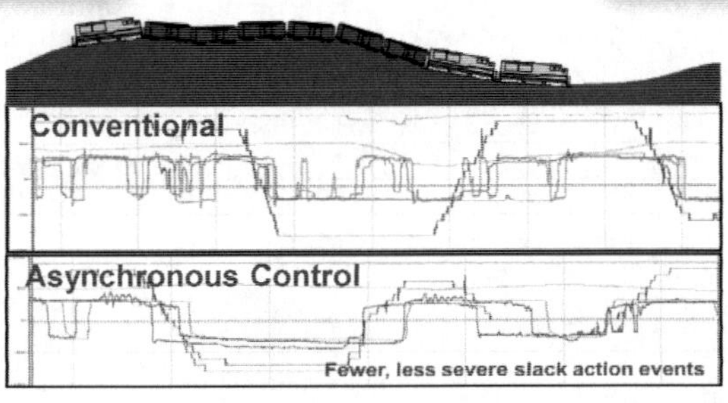

Controlling Slack to Minimize Short/Severe Lateral Force

W/R Interface vs Centripetal Force

Train Driving

Minimal track and gauge impact

Gauge spreading and sleeper impact

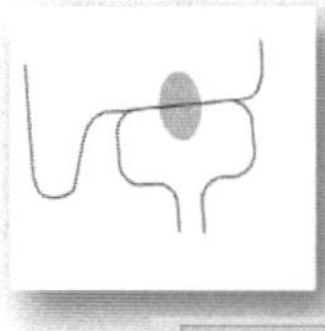

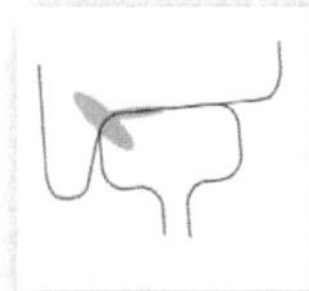

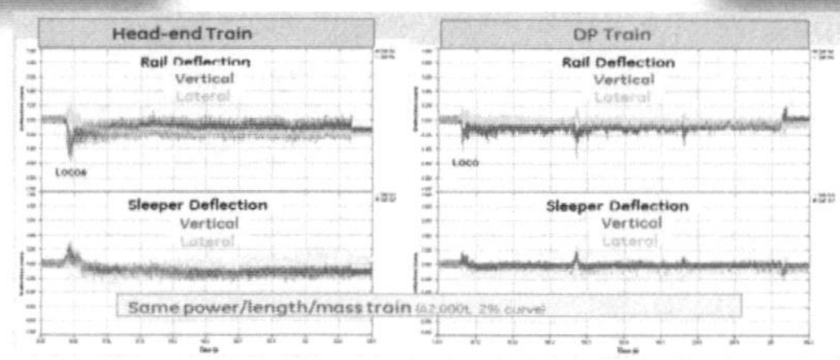

Controlling Train Speed to the Super-Elevation Speed

Evaluate:

Mobile, High Fidelity Equipment is Critical to understanding/Improvement

2001 -2012

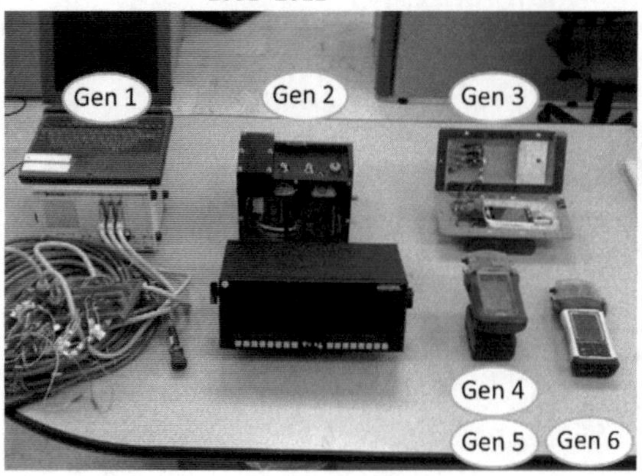

Gen 1: Single location
8ch, 100 S/sec, 8hrs, 15lb

Gen 2: Multi (radio) locations
4ch, 0.5 S/sec, 4hr, 30lb

Gen 3: Multi locations
4ch, 10 S/sec, 6hr, 15lb
(Dell Axim dicontinued)

Gen 4: Multi location waterproof
4ch, 10 S/sec, 10hr, 8lb

Gen 5: Multi location waterproof
4ch, 50 S/sec, 8hr, 8lb

Gen 6: Multi Location waterproof
4ch, 100 S/sec, 96hr 9lb

Human Caused Slack Example

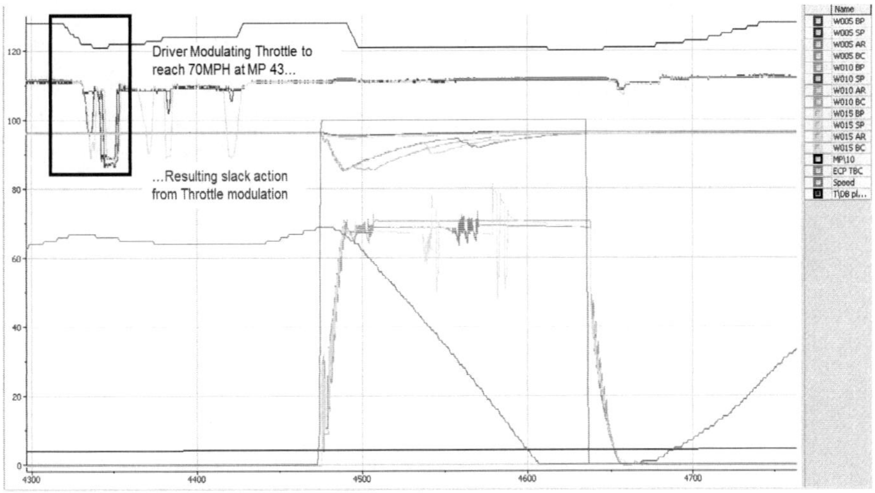

Human Caused Slack Example

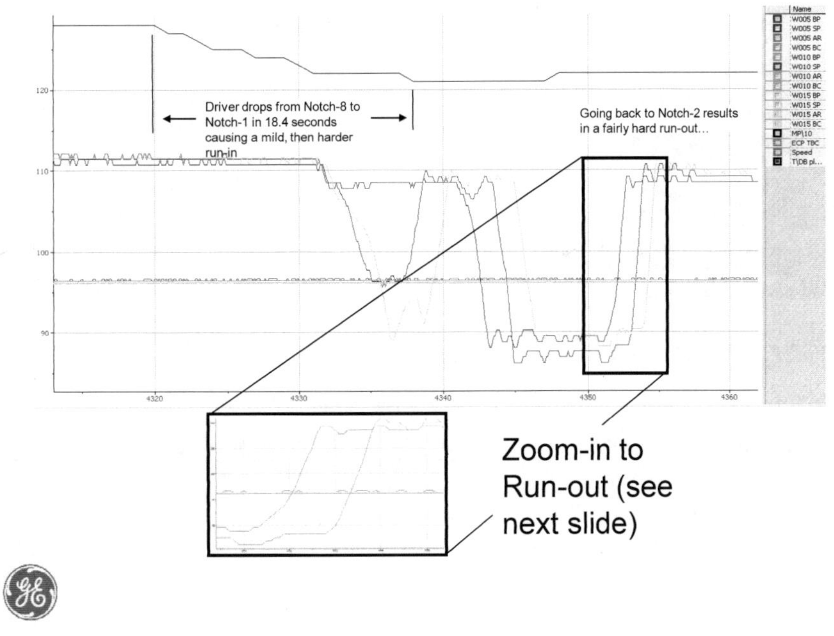

Human Caused Slack Example

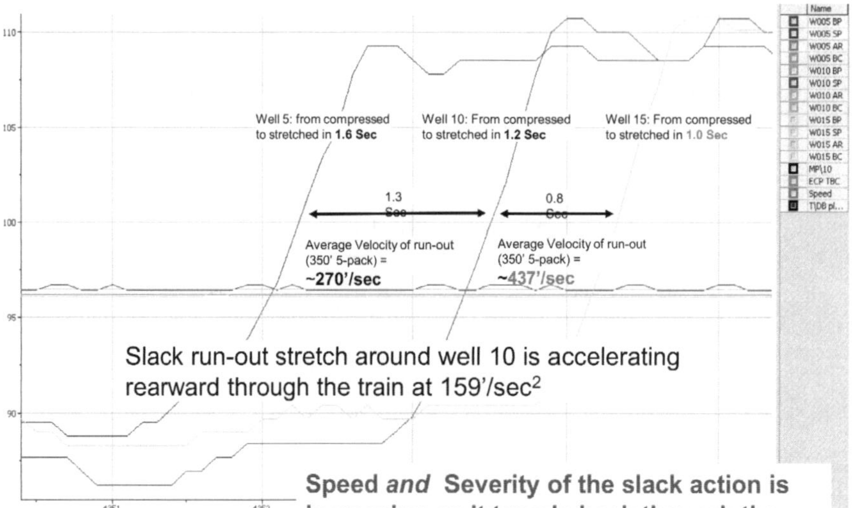

Speed *and* Severity of the slack action is increasing as it travels back through the train

Control: Two Smart "Operations" Products can help control Wheel/Rail Wear

Trip Optimizer:

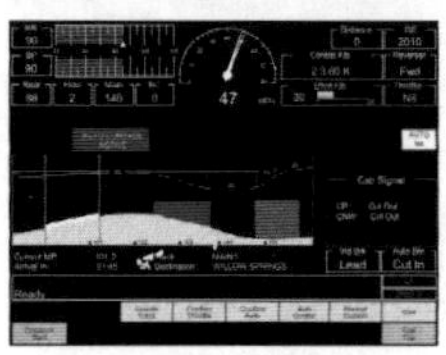

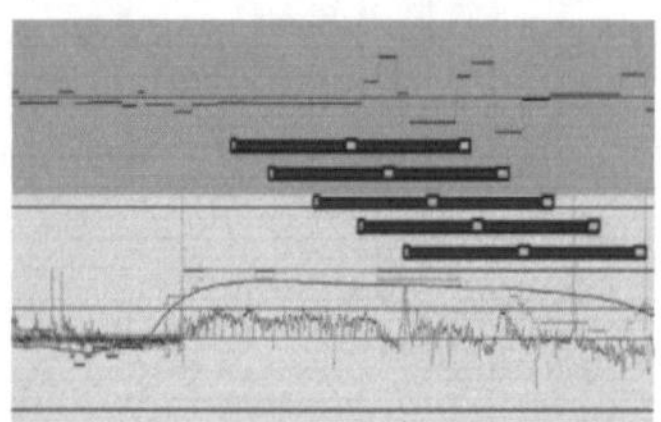

LOCOTROL DP:

- Low Coupler Force:
- Fast Brakes:
  Air Vented at multiple places
- Fast Release:
  Air filled at multiple places

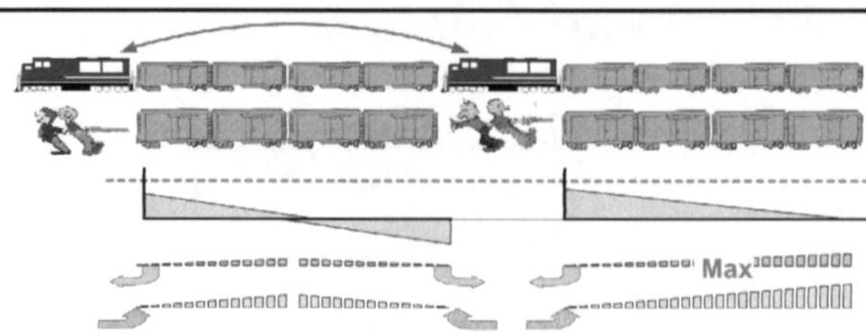

TO Physics- Based Optimization

- **Physics:** There is *only one* optimized trip for a set of constraints.
- **TO does full parametric optimization:** TO can instantly create a family of curves for the entire run for changes to even a single changed or learned parameter:

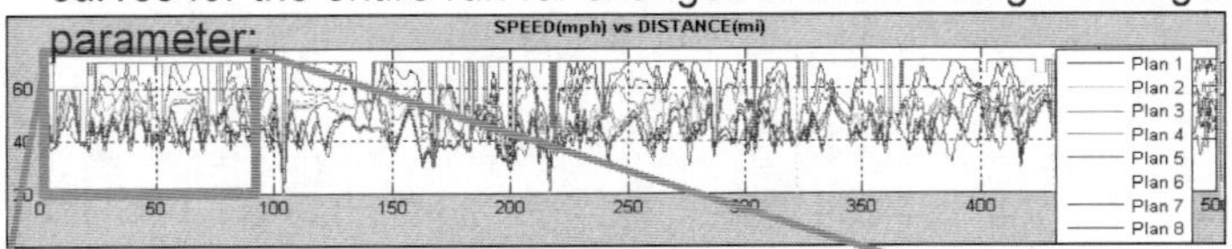

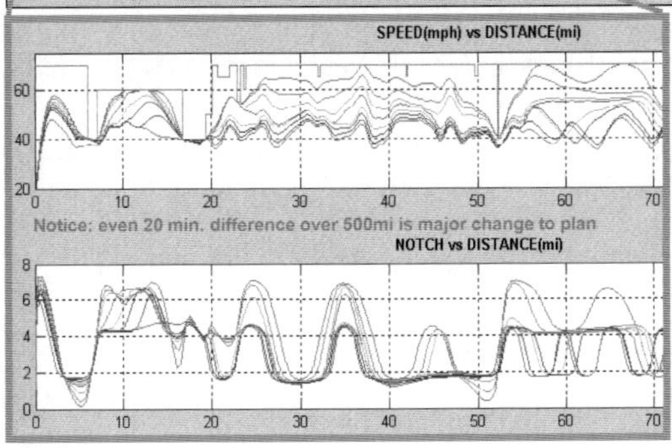

With full physics based optimization, TO can create a family of curves

Trip Optimizer:
Saving fuel by changing just trip time (500mi run)

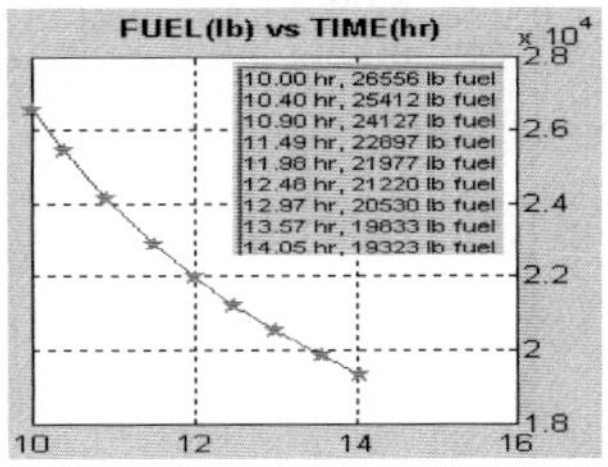

TO Control of Slack Action

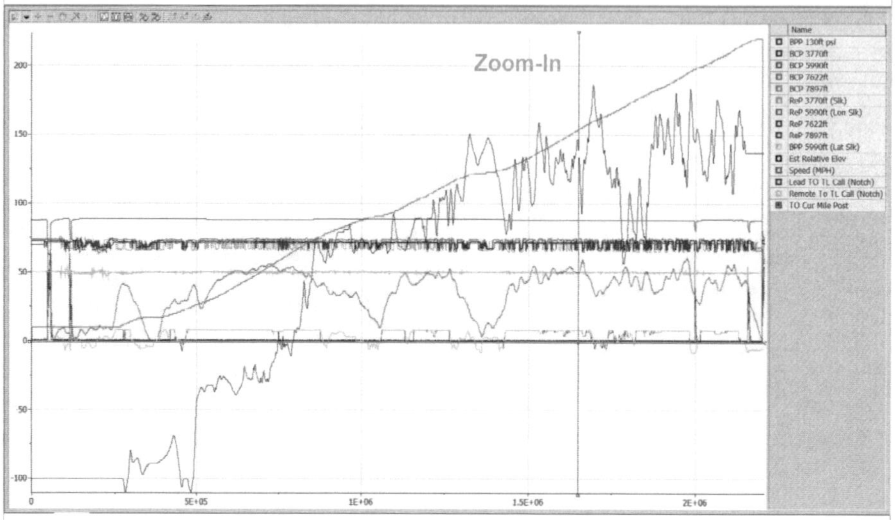

TO Control of Slack Action

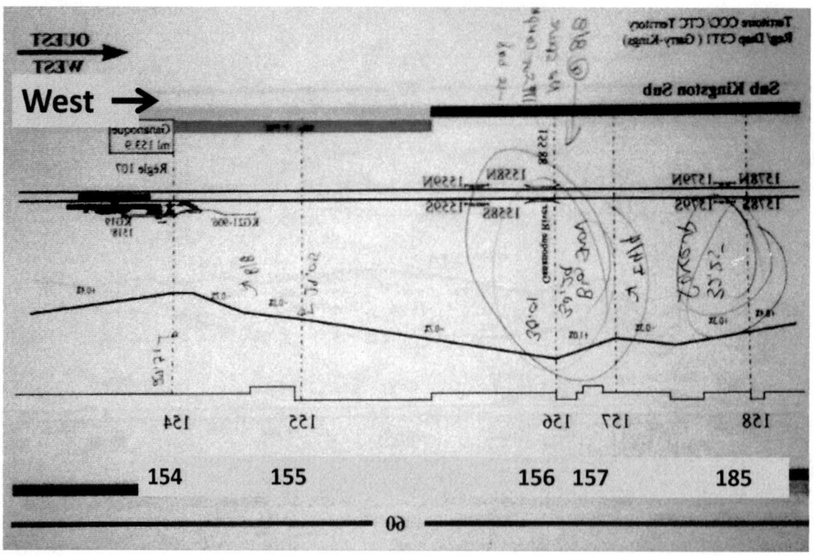

TO Control of Slack Action

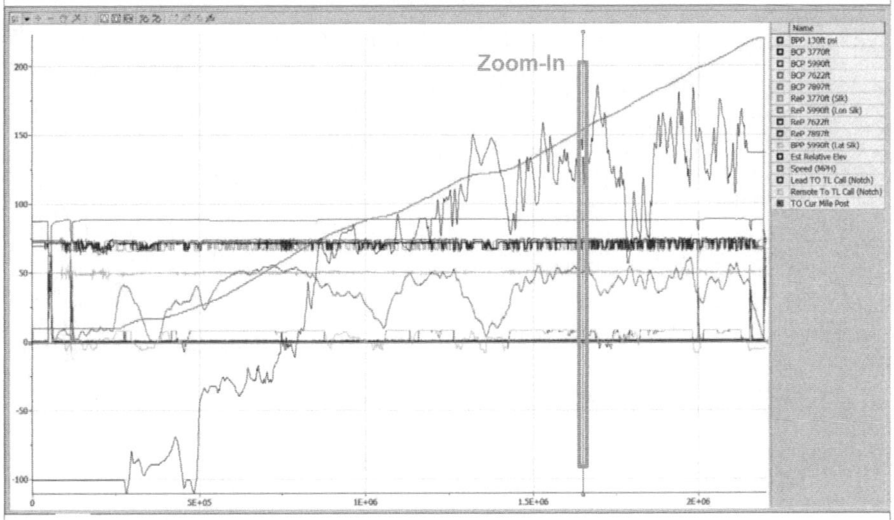

TO Control of Slack Action

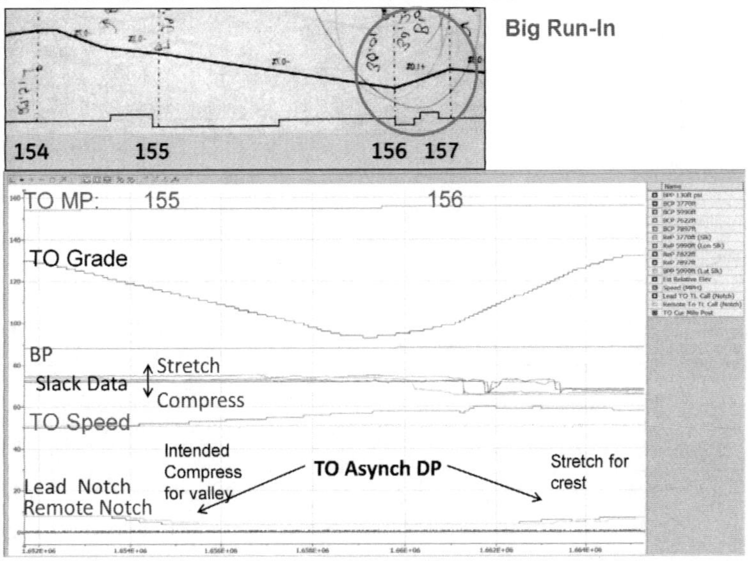

Big Run-In

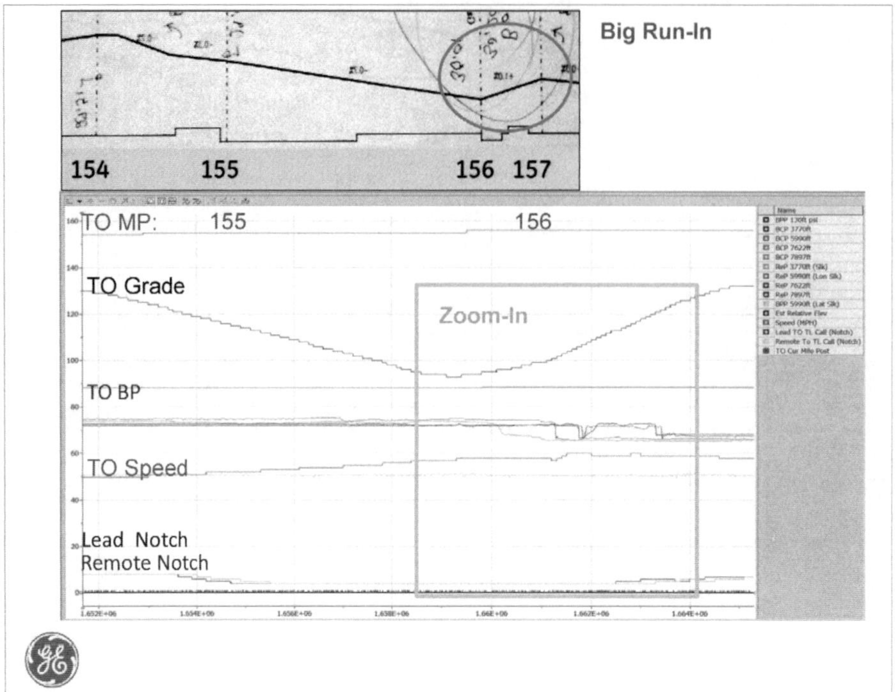

TO Control of Slack Action

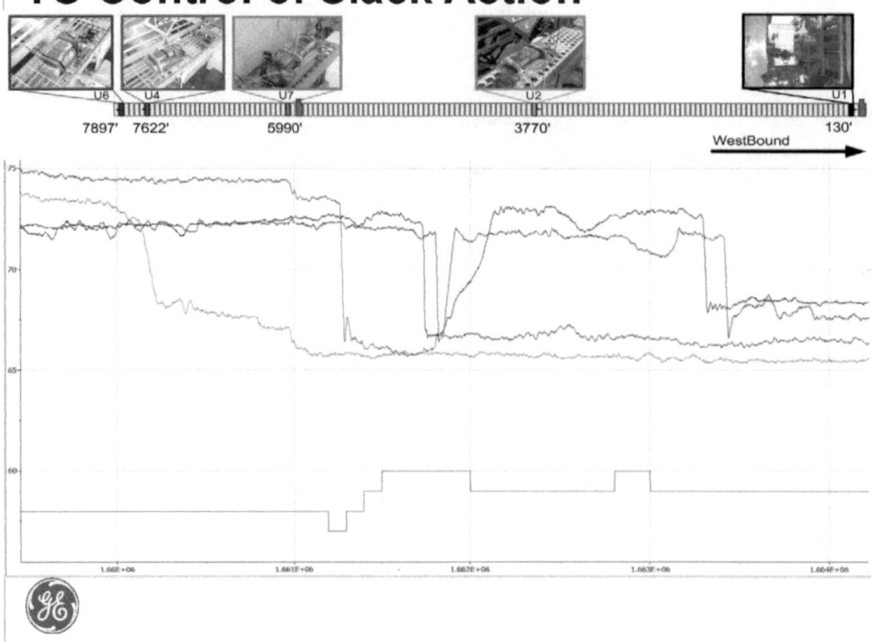

TO Control of Slack Action

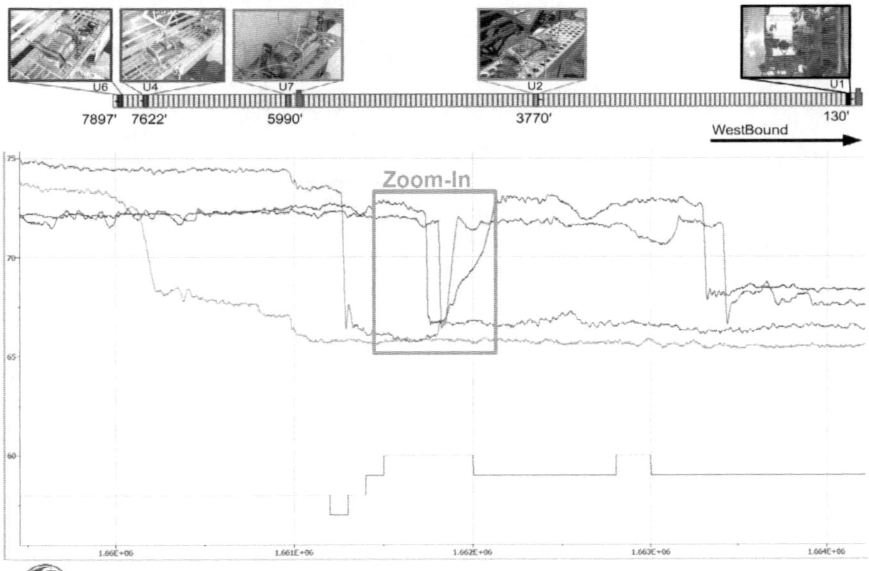

TO Control of Slack Action

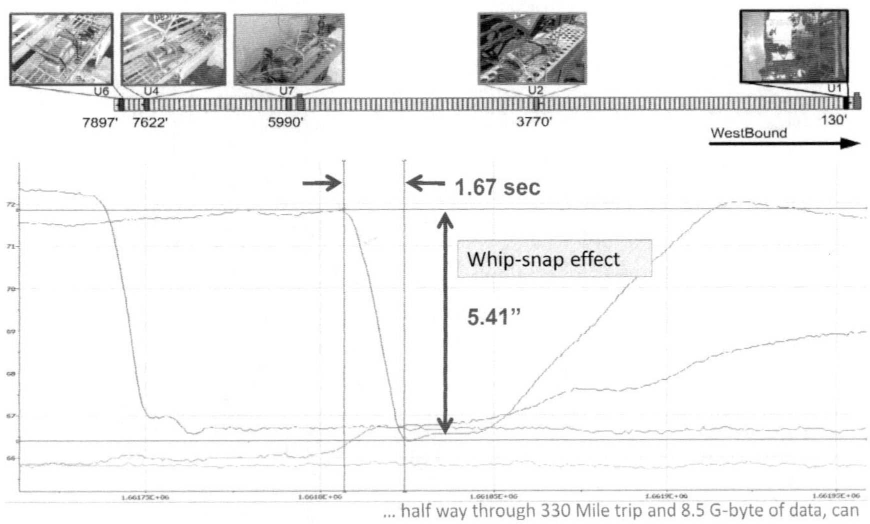

... half way through 330 Mile trip and 8.5 G-byte of data, can measure 1/100" at 1/100th second...

Evaluate: State of the Art Equipment

PPC-Daq R7
4 Channel + GPS
Pressure/0-5V selectable
100 Sample/Sec
86hr logging max
(12hr/battery)

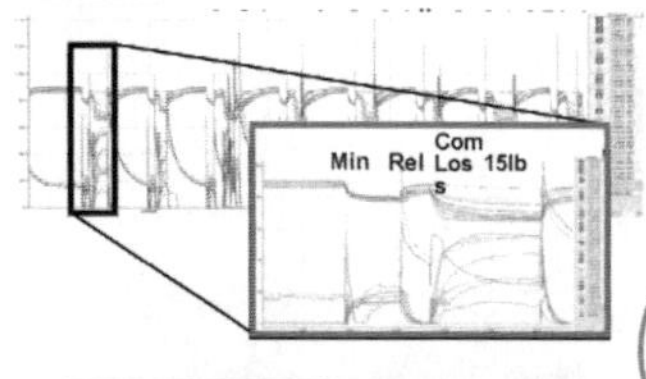

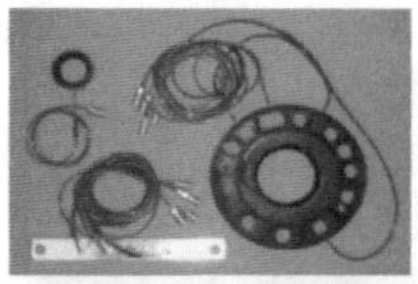

Pressure Interfaces
3/32" tubing/valve quick disconnects
- Glad-hand gasket
- Single Car Test Port
- Service Portion Gasket

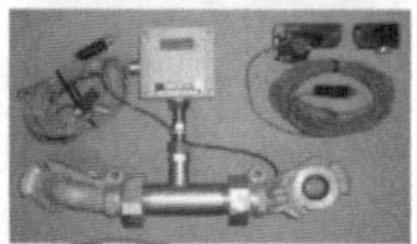

0-5V Interfaces
- 0 to 200 CFM high precision flow meter
- 36" travel string pot for slack measure

NEW!

- Temp: 0.1 Degree f, 100s/s

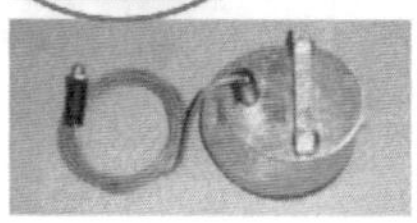

MU 27 pin connector
Converts throttle TLs to stepped voltage for capture of Notch Commands

Detail of Wheel Temperature Sensor

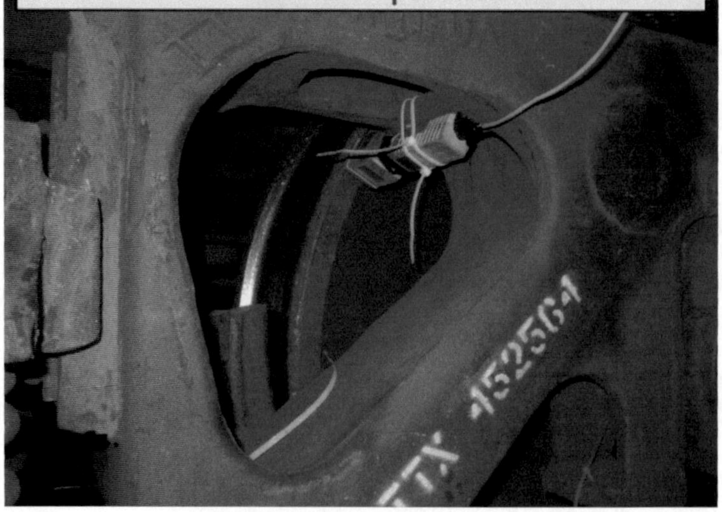

Discovery:

Found that test equipment can detect wheel temperature changes that are directly Related to **curvature and longitudinal train force**

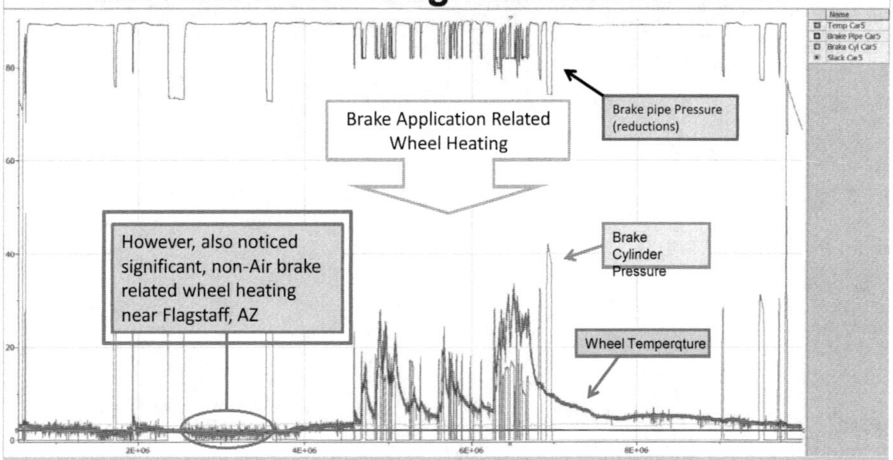

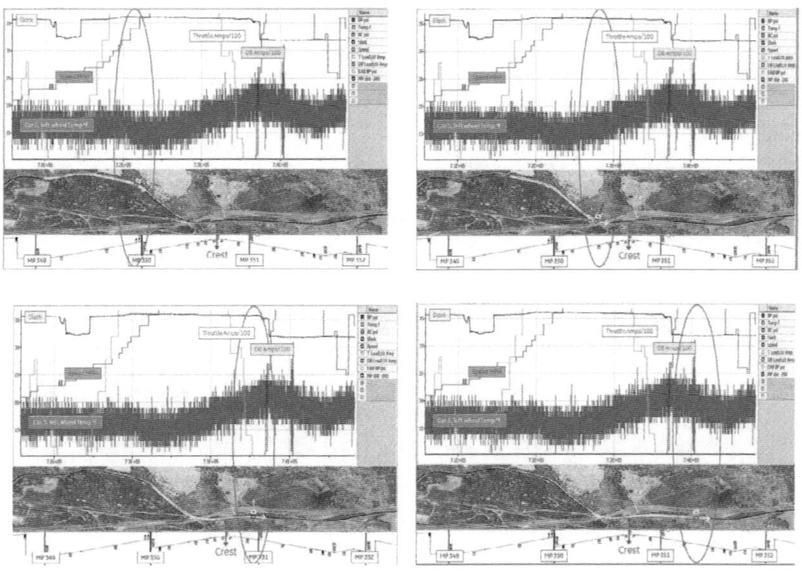

TO and Top Of Rail Lubrication

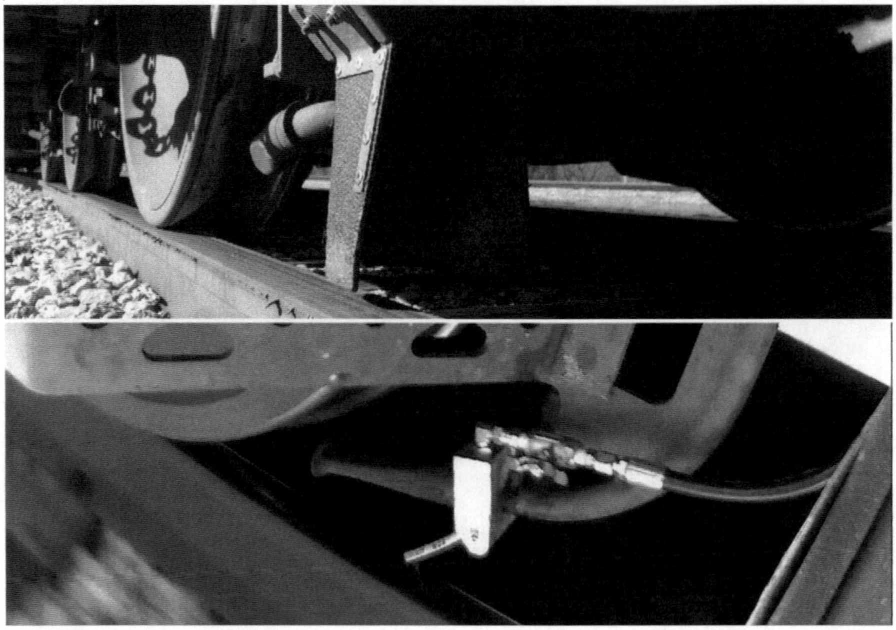

TO Control of Top Of Rail Lube

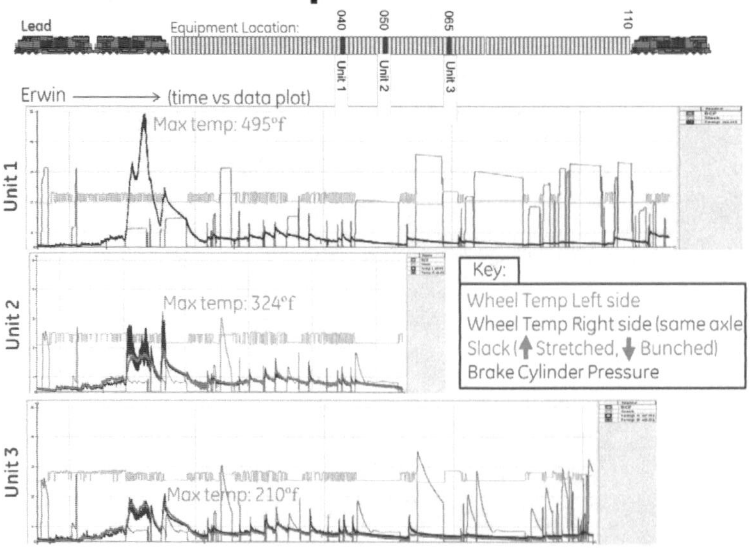

TO Control of Top Of Rail Lube

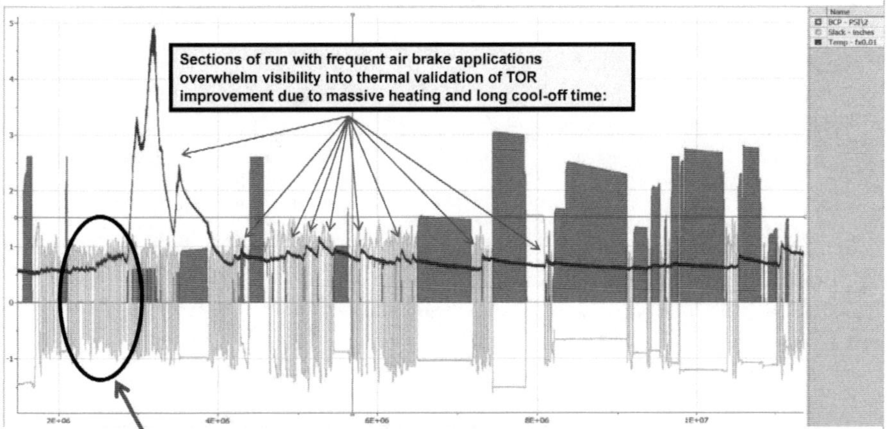

However, long hill climb outside of Erwin appears to show signs of significant difference in wheel heating with TOR on versus off…

TO Control of Top Of Rail Lube

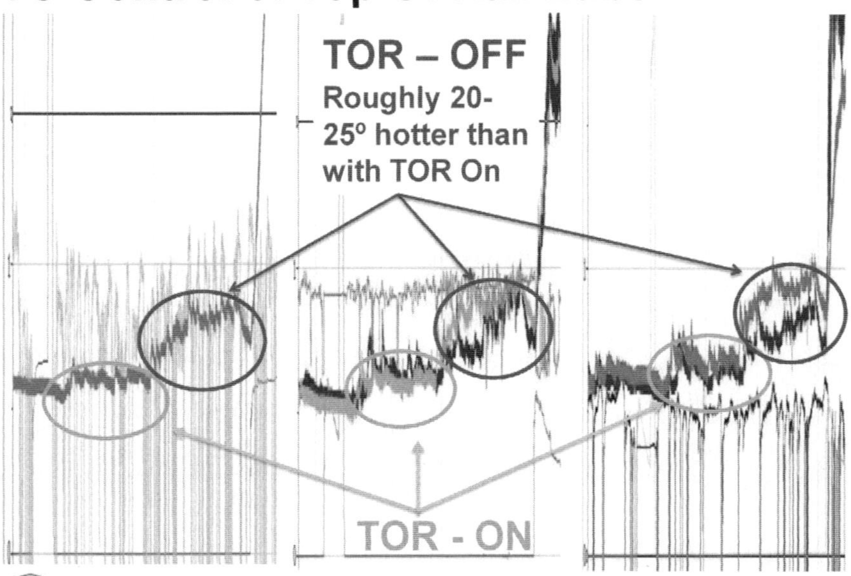

Constantly Improving Capability…